ı about this series at http://www.springer.com/series/13617

Springer Biographi

More informatio

John B. West

Breathing on the Roof of the World

Memoir of a Respiratory Physiologist

Springer

John B. West
School of Medicine
University of California San Diego
La Jolla, California
USA

ISSN 2365-0613 ISSN 2365-0621 (electronic)
Springer Biographies
ISBN 978-1-4939-7121-3 ISBN 978-1-4939-7122-0 (eBook)
DOI 10.1007/978-1-4939-7122-0

Library of Congress Control Number: 2017942788

Cover illustration: Dr. John B. West (Front cover), Dr. Eugene Braunwald (Back cover, top), and Dr. Hermann Rahn (Back cover, bottom)

Printed on acid-free paper

This Springer imprint is published by Springer Nature
The registered company is Springer Science+Business Media LLC
The registered company address is: 233 Spring Street, New York, NY 10013, U.S.A.

Preface

As is clear from the first chapter, I began this Memoir for members of my family, particularly my children and grandchildren. However, as it developed, it occurred to me that it might be of interest to a larger community. I have been fortunate in having had an unusual career. Also, as noted in Chap. 1, I was influenced by recently reading biographies of two medical scientists with whom I previously worked, Charles Fletcher and Archibald Cochrane. I then realized that I was sadly ignorant about so much that was going on in their lives that was related to their research. The result was that I now have a much clearer understanding of their work. Scientific papers are essential of course, but so much else happens that is never related in this medium. I have tried to fill in some of this material here.

The result is this Memoir. The narrative occupies 11 chapters that are amply illustrated with photographs. Many of these deal with my two Himalayan expeditions that I was fortunate to take part in. There are also footnotes to flesh out some of the details in the narrative. This part should be accessible, and hopefully interesting, to many people who do not have a scientific background.

Additional online material is directed at physiologists and others in the academic community who have an interest in respiration and high altitude science. It includes a section that is a commentary on what I think are the most significant scientific contributions. To do this, I have selected a number of articles and included their abstracts that summarize the science and then added comments that hopefully explain why the study is important. I hope that this strategy will make the Memoir useful and interesting to a larger number of people.

I should say something for North American readers about the education process in Australia. This is actually similar to that in the UK and Europe but will be foreign to some US and Canadian readers. The high school that I attended reached a very high level in science in the last 2 years, although as a result, my education was regrettably very narrow. For example in these last 2 years, my subjects were mathematics I, mathematics II, physics, chemistry and, mercifully, English. This last was the only leaven in this otherwise very nerdy selection. As an example of the high level of teaching, the mathematics included substantial advanced calculus.

From high school I moved straight into the Faculty of Medicine at the University of Adelaide. The medical course lasted 6 years followed by the required 1 year of residency. However, I completely missed the college experience that all North American students are exposed to. I regret this. Our children, who were both born in the USA, went to college of course, and I became very aware of the advantages of this broad education. I think this system is superior to the one I was exposed to, although of course it is expensive. Incidentally, some of the medical schools in Australia and the UK are now moving to a pattern more like that of the USA.

Another disadvantage of my education is that I was only 17 years old when I entered medical school. This is far too young. I now teach first-year medical students who are about 22 years old. These 4 or 5 years make an enormous difference at that age, and the students are far more mature and receptive than I was. Remarkably, I was only 23 when I received my medical degree, and a year later I was qualified to remove your appendix if you were brave enough to let me.

Many people have helped with this Memoir. Amy Clay produced an early draft 14 years ago. Lisa Richter has been invaluable with the present text, and Zhenxing Fu has been indefatigable with the figures. My wife, Penelope, has been a continual tower of strength. I am greatly indebted to all.

La Jolla, CA, USA John B. West

Contents

1 **Early Days, 1928–1945** 1

2 **Adelaide University and Medical Residency, 1946–1953** 13

3 **London and Llandough, 1953–1956** 19

4 **Postgraduate Medical School, 1956–1960** 33

5 **Silver Hut Expedition, 1960–1961** 45

6 **University of Buffalo and the Postgraduate Medical School, 1961–1967** 61

7 **Palo Alto and the NASA Ames Research Center** 71

8 **University of California, San Diego, 1969–1981** 79

9 **American Medical Research Expedition to Everest, 1981** 89

10 **Studies of Astronauts During Spaceflight** 109

11 **Other Research Projects** 119
- 11.1 Stress Failure of Pulmonary Capillaries 119
- 11.2 The Avian Lung 122
- 11.3 The Elephant Lung 125
- 11.4 Archive on High Altitude Medicine and Physiology 127
- 11.5 Oxygen Conditioning 128
- 11.6 Assessing Pulmonary Gas Exchange in Patients by Analyzing Expired Gas 130
- 11.7 Pulmonary Function in Pilots of High-Performance Fighter Aircraft 130

12 **Commentary on Some of My Books and Scientific Papers** 133

Chapter 1
Early Days, 1928–1945

Why a memoir? Vanity of course. But seriously, there is another reason. I have had an interesting life (and for the last 50 years have shared it with Penelope) and it seems wrong if our children, and perhaps their children, do not know something about it. History is often fascinating, and, for example, I very much regret that I know so little about many aspects of my father's life. Of course he told us some anecdotes of his boyhood but I would love to know more about his period at Adelaide University while studying medicine, his training at the Royal Adelaide Hospital, his early days with my mother, his early experiences in general practice in Jamestown in the 1920s, his periods of training in Edinburgh and Liverpool, his experiences during the war, and his subsequent very successful career as an orthopedic surgeon. Apart from a few isolated facts, I know little about what must have been a very interesting life and certainly there is nowhere to turn to read about it. My account should ensure that our children, and perhaps their children, do not have the same regrets.

There is another reason for this memoir, and this may resonate more with many readers. I have been involved with many scientific projects, and some of these have been unusually interesting, for example the Silver Hut Himalayan expedition in 1960, the Everest expedition in 1981, and our more recent studies of astronauts in space. All of the results of these studies have been written up in papers in scientific journals.[1] However these formal accounts omit much of what was actually going on during these projects. An informal memoir like this allows the opportunity of "telling it like it was".

I became more aware of the interest of this anecdotal information when I recently read two autobiographies of prominent clinical scientists with whom I have worked. One was Charles Fletcher[2] who, as will become clear in Chap. 3, was

[1]My curriculum vitae and a full list of books and articles can be found in doi:10.1007/978-1-4939-7122-0_12.

[2]Pioneering physician: the life of Charles Fletcher 1911–1995 by Max Blythe and Richard Peto. Bicester UK, words by design, 2016.

J.B. West, *Breathing on the Roof of the World*, Springer Biographies,
DOI 10.1007/978-1-4939-7122-0_1

largely responsible for me choosing respiratory physiology as a career. He also set up the Pneumoconiosis Research Unit in which I spent a year in 1955, and this was one of the best years of my life. However when I read his biography, I was taken aback to learn that Fletcher had serious reservations about some parts of the physiology program that seemed so innovative to me.

The other autobiography was by Archibald Cochrane[3] whose name will be familiar to some readers because of the Cochrane Database. His influence on my career was not nearly as important as that of Fletcher. However his account of the early days of epidemiology of which he was one of the founders, opened my eyes to the unevenness of scientific progress. Incidentally, readers who are more interested in science than my early days might like to skip parts of the first couple of chapters of this Memoir.

I was born on December 27, 1928 at the Memorial Hospital in Adelaide. I still have an official copy of my birth certificate, and one oddity is that my mother's first name is given as Neta whereas her correct name was Meta. Presumably the error occurred because the name is so unusual and it was misheard. My sister Rosemary was born 2 years after me on December 1, 1930.

At this point I should say something about my parents. My father, Esmond Frank West (Fig. 1.1) was born on March 31, 1899. Parenthetically the 18 often irritated him because all the official forms he had to fill out had the date beginning with 19… and he said this made him feel antiquated. Of course 1899 was indeed a long time ago and Queen Victoria was still on the British throne. My father came from a large family of nine children (Fig. 1.2) and was the eighth child of William Arnold West (1856–1922) and Emily Burnard (1856–1933). William was a teacher who did well to become Inspector of Schools in South Australia, but the large family had modest means. Rosemary remembers that my father often spoke of his impoverished childhood. They lived on jam made from home grown plums and apricots boiled in the "copper" (usually used for washing clothes). He talked of having "bread and point" for a meal—that is when they were particularly short, the younger children would be given bread and dripping and then point to what they would really like to have on the bread, but not be allowed to eat it! Nevertheless grandfather West was remarkable in that seven of the nine children had a tertiary education at a time when this was unusual. Only Lorna (who left school relatively early to help care for the large family) and Arnold whose health was poor did not have a tertiary qualification. Even the other two daughters did when few girls went to university. The family lived in a typical, attractive, bluestone house named Arlington at 49 Hewitt Avenue, Rose Park, a suburb of Adelaide. The house was built in 1894 and still stands. My father's forebears were researched by my uncle Reginald and I have an extensive family tree.

[3]One man's medicine: an autobiography of Professor Archie Cochrane by Max Blythe and Archibald Cochrane. London: British medical journal, 1989.

Fig. 1.1 My father, Esmond Frank West (1899–1988)

Fig. 1.2 My father's parents and their family of six sons and three daughters. Back row: Arnold, Stanley, Roy, Marjorie. Middle row: Reginald, my grandfather, my grandmother. Front row: Lorna, Ruth, Esmond, Eric

My father was educated at Norwood Primary School and Norwood High School, and then Adelaide High School. This is a state school with little or no tuition fees and perhaps this was one reason why it was chosen. However my father's much older brother, Reginald was a teacher at the school (he subsequently became principal) and that was probably the main reason. Adelaide High School probably did not have the social cachet of the large private schools such as Saint Peter's College and Prince Alfred College but it provided an excellent education. For example, my father had a life-long love of English literature and could quote Shakespeare at length. His knowledge of the English classics was nurtured by his father who used to read a chapter of a Charles Dickens novel to the whole family every evening. My father studied medicine at the University of Adelaide and graduated M.B.B.S. in November 1922. A curriculum vitae that he prepared states that he took the top place in the last 4 years of his course. It is interesting that the year before my father's included Howard Florey (1898–1968) who later became a Nobel Laureate (1945) for his part in the discovery of penicillin.

Rosemary and I do not know much about his period as a medical student but there are some anecdotes about the professor of anatomy, Archibald Watson. He was an eccentric and used to ride his motorbike around the dissecting tables in the anatomy room. According to one account[4] there was a price on his head at one time because he was a crewmember of a pirate ship that kidnapped natives in the Pacific islands and subsequently sold them. Another less colorful story was that he was invited to a somewhat formal weekend party at a grand house in the Adelaide Hills but arrived without any luggage. When reminded that it was customary to dress for dinner he replied that he was wearing all the appropriate clothes under his day suit together with his pajamas, *etc*.

After graduation father spent a year as a resident at the Royal Adelaide Hospital and during this time (or before) he met my mother who was a nursing sister. They married on October 18, 1924. We know little about his residency except that he was involved in the very early use of insulin for the treatment of diabetes in Australia. This came about because T. Brailsford Robertson (1884–1930), who was the professor of physiology and biochemistry, had recently returned from Toronto where he had worked with the Banting and Best group who had discovered insulin, and with Macleod had received a Nobel Prize in 1923. Brailsford Robertson had learned how to extract insulin from animal pancreas, and apparently my father was involved in treating some of the early patients by injecting this new and presumably crude material.[5] Of course diabetes was a terrible scourge until insulin became available. Following his year of residency at the Hospital, my father went into general practice in Jamestown, a small country town about 200 km north of Adelaide.

My mother, Meta Pauline Spehr (Fig. 1.3) was born on October 13, 1897 in Millicent, a prosperous farming town in the southeast of South Australia. Her father, Georg August Spehr (1853–1925), had emigrated from the Hanover region of Germany and married Louise Pauline Pohlner (1862–1922). Louise's grandfather, Ignatz Pohlner (1787–1871), had emigrated from Germany in 1847. Both the Spehr and Pohlner families have done very extensive genealogical research. There

---

[4]WGK Dawson and RA Leonard. The University of Adelaide 1874–1974 Adelaide, Rigby, 1973.

[5]According to the entry in the Australian dictionary of biography (edited by G. Serle, Melbourne University press, 1988) Brailsford Robertson made an important contribution in improving the method for preparing insulin. It was manufactured for the first time in Australia in the Darling building on the Adelaide University campus.

Fig. 1.3 My mother, Meta Pauline Spehr (1897–1995)

are books on both families[6,7] and the Spehr family book has a hard cover with over 470 pages and many fine photographs. My mother was one of ten children (Fig. 1.4) and was brought up on a large farm named Willowbank three or four miles outside Millicent where the schools were situated. She always walked the three miles to school, sometimes over muddy roads, and according to Rosemary who had good opportunities to hear her stories, she sometimes spent the whole day there in wet shoes. Many of her relatives remained in the Millicent area and Rosemary and I used to spend the summer holidays with one of the families. One of my memories is of my mother undoing the stitching of my jacket, putting a £1 note inside and stitching it up again. This was apparently my keep for the summer! I suppose we had good and bad times at Millicent but I tend to remember the bad ones, for example, when my cousins of about my age taunted me into trying to lift bags of wheat which I could not do.

[6]The Spörer, Speher, Spoehr and Spehr family in Australia, 1847–1986. Spehr Family Reunion Committee 1986, Adelaide: Lutheran publishing house, 1986.

[7]The family history of Ignatz Pohlner and his Wife Johanne Elenore nee Wunderlich and their known descendants. Anonymous, Adelaide: Lutheran publishing house, undated.

Fig. 1.4 My mother's parents and their family at Willowbank. Left-to-right: Myrtle, my mother, my grandfather Georg, Eda, my grandmother Louise, Anne, Edward, Hilda, George, Walter, Eileen (*in front*). Henry was not yet born

My mother presumably left school at the age of about 14. She was a beautiful young woman (the Spehr girls had a reputation for good looks) and an accomplished horseback rider. Later she moved to Adelaide for training as a nurse at the Royal Adelaide Hospital where she excelled and received a gold medal as top of her class. She told a number of stories about this period to Rosemary. Times were hard. She used to work 12 h shifts with 1 day off a week and there were no wages for the first 3 months—just "keep" plus uniforms and perhaps a little pocket money. She was called a "black pro" for this period because the uniforms were black. But there were amusing anecdotes. If they had a death in the ward just before the end of the night shift, they would prop the body up with a cup of tea and pretend that he had only died when the day staff came in! Otherwise they would have to work an extra hour or so laying the body out. They also had to dismantle and scrub the iron bedsteads after each patient left to get rid of the bed bugs.

My parents lived in Jamestown for a short period after their marriage. The only information that I have about this period is my father's anecdote that my mother had to run fast to round up the cow. Presumably they kept one for milk. However in 1925 my father went to the U.K. for advanced surgical training where he received the degree F.R.C.S., Edinburgh in March 1926. I know that my mother joined him for part of the time at least. My father used to say that he had never been so cold as he was during his period in digs in Edinburgh and it took him several years to warm up!

When he returned to Adelaide he joined Dr. Frank Beare in general practice at Mile End, a gritty industrial suburb on the west side of Adelaide. My father handled the surgical side of the practice while Frank Beare (known as Teddy Beare) concentrated on internal medicine. There is an anecdote about his return to Adelaide but I am vague about the details. Apparently he did not have any plans for where to work but someone told him that Frank Beare was looking for a partner. Father telephoned him and the agreement was made almost immediately. During this time my parents lived in Rowland Road, Hilton in a large house with a garden and tennis court. My father used to enjoy tennis, golf and bowls. The house no longer exists and the area has been redeveloped. I have almost no memories of the house except that there was a sandbox, and I had a tricycle. My father used to relate that at one time my tricycle was stuck somewhere in the garden and someone offered to help but I retorted that I would "do it my sauce."

In 1934 when I was five, my father returned to the U.K., this time to Liverpool, to study for the degree of Master of Orthopaedic Surgery, M.Ch. Orth. Liverpool was an important center for orthopedic surgery at that time partly because it was a large port and ships would arrive with sailors with terrible bone injuries having fallen off a mast or whatnot. There were famous orthopedic surgeons in Liverpool including McMurray and Reginald Watson-Jones. The latter was a very colorful surgeon and wrote an influential textbook on orthopedic surgery. I remember him coming to our home in Adelaide for dinner many years later. My parents and I also went to a party at his elegant flat in Portland Place in London when they visited me in the early 1960s. The flat was near the statue of Lord Lister in the middle of Portland Place and was unusual in that it contained a pipe organ that Watson-Jones played enthusiastically. There was prodigious drinking and my parents and I subsequently went to France and Germany for a tour in my car where we gradually recovered from the party and would refer to the date as Watson-Jones plus 1, plus 2, *etc.* Watson-Jones was president of the British Orthopaedic Association when my father was president of the Australian Orthopaedic Association, and there was a meeting in London in June 1952 when my father was presented to Queen Elizabeth, the Queen Mother together with Watson-Jones. My father received his orthopedic degree in November 1935 and then spent a short time with the famous orthopedic surgeon Lorenz Böhler in Vienna. He returned to Adelaide in February 1936 to start full time orthopedic practice both at the Adelaide Children's Hospital and as an assistant to the well-known surgeon Lionel Betts at the Royal Adelaide Hospital.

While my father was in Liverpool, my mother and the two children lived in a small rented house in Scott Street, Dulwich. I suppose the large house in Rowland Road was sold. The main advantage of this house was that it was close to 14 Gurney Road, Rose Park where the Potts family lived. Lorna was an older sister of my father's who married Gilbert (Gib) Potts and according to Rosemary, Lorna was a second mother to my father when he was young. The large Potts family included Renfrey (Ren), Don, Helen, Dorothy and other sisters who were older. Helen and I were about the same age and used to play together sometimes. Ren was 2 or 3 years older but we became close friends and have been so all our lives. Ren was a distinguished applied mathematician who was a Rhodes Scholar and subsequently returned to Adelaide from Oxford where he became the Chair of the Department of Applied Mathematics and a very influential person in the University of Adelaide. He received the Order of Australia (AO) and was one of my closest links with Adelaide.

Another advantage of the location of the rented house was that it was close to Rose Park Primary School which was only about 10 min walk away. Father's youngest sister, Ruth, was a teacher there. My only memories of the school are the drum and fife band in which I played the fife and I recall that "Men of Harlech" was part of the repertoire. My memories of the Scott Street house are very vague although I do remember getting lost on one of the nearby streets on my tricycle and giving my address as Scott Street, only to find that there was another street of the same name in an adjacent suburb.

I must have been at Rose Park Primary School from 1934–1937 when I was enrolled in the Preparatory School at Prince Alfred College, presumably in February 1937 at the age of eight. It was natural for me to go to P.A.C. because Uncle Gib was one of the most prominent teachers there and he eventually became Assistant Headmaster. Also P.A.C. was the Methodist high school and had a very high standard. However the most prestigious high school was the Anglican Saint Peter's College but this was in North Adelaide some distance away. There was always, and still is, keen but friendly rivalry between Prince's and Saint's.

When my father returned from the U.K. in early 1936 he bought 11 Fergusson Square, Toorak Gardens This was in the same area of east Adelaide as Dulwich but much nicer. Fergusson Square is a very attractive location having a large public garden in the center and no through traffic. However the house was relatively modest with small front and back gardens and unlike the house in Rowland Road it had no tennis court. I have often wondered why my father who had a very successful practice in orthopedic surgery did not move into a larger house at some stage. It is true that Fergusson Square was well situated being only 10 min drive to the center of Adelaide. I used to go to P.A.C. on my bicycle and that probably did not take more than 15 min or so.

My father was very fond of the land (rural country) and according to my sister spent many of his Christmas holidays working on a farm in Yorke Peninsula west of Adelaide helping with the harvest, and sowing wheat bags by hand. Later in life he bought a large property in a fairly remote part of South Australia near Coonalpyn. The soil lacked trace elements but when these were added with the super-phosphate fertilizer it became more productive. There were tax advantages in developing this land the locals used to refer to city people who did this as "Rundle Street farmers" (Rundle Street is one of the main streets in Adelaide). My father was in partnership with Dr. Sam Pearlman, an ophthalmologist, whom he liked very much. Father spent many weekends down on the farm and I went down on several occasions. It was only a 2 or 3 h drive from Adelaide.

I continued to live in the Fergusson Square house until I left Adelaide, and indeed my parents stayed there until they were too old to look after themselves.

They then moved to a "granny flat" adjacent to Rosemary's house in Chiltern near Beechworth in northeastern Victoria. The Adelaide house was then sold by auction and has since been lavishly remodeled with a swimming pool, *etc*. The house was generally comfortable although I lived in a "sleep-out" which was an outside room protected by a wire screen. One disadvantage of the house was that the front faced west and in summer it used to become extremely hot with the fierce afternoon sun. Air conditioning was not introduced until after I left.

I entered P.A.C. in form II B and came second in the class as evidenced by the book as second prize that I still have. It is beautifully bound in crimson with the school crest on the front cover. I also received second prize in form II A in 1938, third prize in III B in 1939, and first prize (Dux) in III A in 1940, the last year of Prep school when I was 12, about to turn 13. According to Rose there was some excitement when I did my final exams because I had some childhood illness and my mother at least (she was the ambitious one) wanted me to be Dux. Accordingly a Methodist minister came to the home to act as a proctor! I have few memories of my prep school although I did spend one term as a boarder and was thoroughly miserable. I do not know why my parents tried that experiment.

Of course it was a big step to go to the Upper School (which we called the Big School) although this was just across the road. Our class IV B was joined by a number of new pupils who had been to prep school elsewhere. They included Peter Stobie and Bob Steele and I still vaguely remember seeing them for the first time because they came with a reputation as strong students. We stayed together during the whole of the Upper School and generally our class was very strong academically. I still keep in touch with Peter and Bob and indeed they were with me through university. The forms in the Upper School were IV B, IV A, V B, V A, VI A and Upper Sixth. Happily this strange classification which of course came from the U.K. has now been replaced by years 7, 8, *etc*. I was generally a conscientious student and did well although Peter Stobie or Bob Steele generally came top of the class. As was often the practice then, I spent 2 years in the Upper Sixth form and left P.A.C. in December 1945. There were probably two reasons for this. First I was still fairly young being only 16 (about to turn 17) when I left school. In addition it was common for good students to spend 2 years in the top form in the hope of obtaining a State bursary (scholarship) because this paid your way through Adelaide University. In fact eight students from our final year class obtained state bursaries which was a very strong showing. There is an extensive account of the performance of our class in the P.A.C. *Chronicle* of January 1946 that I edited.

I had mixed feelings about school like everyone else. On the positive side I was greatly influenced by our physics and chemistry master, Ray T. Smith, whose charismatic way of discussing scientific problems has stayed with me all my life. He suggested that we prepare a P.A.C. *Science Journal* (Fig. 1.5) and I was the first editor. It is now a very glossy production and 2017 is its 73rd year! It is sent to me every year and is a colorful reminder of how long ago I left school. When I visited P.A.C. in 2001 and 2015 we had brief celebrations of its success and I met the boys who were working on the current issue. I wonder how many other high schools produce anything as ambitious and enduring as this. Another master who influenced

Fig. 1.5 An issue of the PAC *Science Journal*. I was its first editor

me greatly was Alan Dennis who taught me to enjoy English literature. During my last year at school I was a prefect and, as was the custom then, one of our privileges was to cane boys for misdemeanors such as smoking. This now sounds distinctly Dickensian!

Outside the classroom I was very active with the scout movement and became an Eagle Scout. We had a cabin at Mylor in the Adelaide Hills. I played cricket and was a reasonably good leg break bowler and batsman but never made the first eleven. I did not like the roughness of Australian Rules football. Unfortunately I was not chosen for the tennis squad which was a pity because the coach, T.G. (Tagger) Luke, produced some outstanding players. I did compete in intercollegiate athletics, coming third in the mile. I was also active in the Debating Society. The majority of students at P.A.C. were day boys from Adelaide. The boarders were from rural areas of South Australia and tended to have a different outlook because most of them were planning to spend their life on the land.

Unfortunately we had a poor headmaster when I was at P.A.C. John F. Ward was apparently a good classical scholar but he was quick to lose his temper and was not a good role model. He flew into a rage if he found a boy not wearing his cap, or riding a bicycle with his coat unbuttoned (this happened to me). On one occasion someone complained that their sand-shoes (sneakers) were missing and he lined up the whole school of several hundred boys on the front oval to have their shoes inspected!

P.A.C. has always had a good academic reputation and has recently gone from strength to strength so that is now one of the better high schools in Australia. I always enjoy reading the annual *Chronicle* which is a very glossy production. Penelope and I visited the school in 2001 and I was honored as one of the first 50 or so "Princes Men," and gave my lecture on our 1981 Everest Expedition to the whole Upper School.

I had several hobbies at home during my period at school. One was making small balsa model airplanes, chiefly gliders, and at one stage there was a small group of us who formed a club. Oddly enough flying radio-controlled model gliders has become one of my most enjoyable hobbies now. In those days there was a magazine called *Flying Aces* that I read avidly. Another hobby was making radios. The first was a crystal set but I later made a valve radio. The most influential magazine in this area was *Radio & Hobbies*. In my later years at school I constructed more sophisticated electrical and electronic devices including one to produce wave patterns in water. When I look back on it there must have been some danger of electrocution from the 240 V supply! I also constructed a small chemistry laboratory in my sleep-out and bored a hole through the window frame separating this room from the kitchen. This enabled me to link up with one of the gas ports on the kitchen stove so I could supply gas to my Bunsen burner. My father had old chemistry text books and I spent a lot of time reading these. There was no doubt that I wanted to go into some area of science.

On Sundays I went to Sunday School when I was young and later to church at Gartrell Memorial, the local Methodist church. The music was very fine in the Victorian Methodist tradition and it gave me a very good sense of harmony. The congregation used to sing lustily in four parts. Each year there was an anniversary celebration for the Sunday School and this involved assembling a large multilevel stage in the church and several weeks of rehearsals. But apart from Sunday, P.A.C. was very much the center of my activities during these years and we did little as a family together. This was partly because with the outbreak of war in September 1939, my father transferred from the militia (army reserve) to the regular army and we consequently saw little of him. In June 1940 he helped to organize the Second 8th Field Ambulance, and he was posted to the Middle East in 1941. He spent much of his time in Palestine where he treated men who were wounded in the North African campaign that involved many Australians. One of his anecdotes was about some donkeys that the hospital had bought from a local Arab for some reason. After a few days the animals that were normally docile became so wild that they were uncontrollable. When complaints were made to the vendor he asked what they had been doing to the animals and the response was that they had simply fed and watered them. "You fed them," he said, "Never do that!"

My father returned to Australia in 1943 but then spent a period at army hospitals in Ballarat and Heidelberg in Victoria eventually reaching the rank of Lieutenant Colonel. Consequently I saw very little of him during my last 5 years of schooling. I also saw relatively little of Rosemary who attended another good private high school, the Presbyterian Ladies College. She had her own friends and there was little social interaction with mine.

Chapter 2
Adelaide University and Medical Residency, 1946–1953

I entered the Faculty of Medicine, University of Adelaide (Fig. 2.1) in February 1946. As indicated earlier, I did well in my final year of school (which I took twice) and according to the P.A.C. *Chronicle* of 1946 I came tenth in the state of South Australia. Bob Steele was first and Peter Stobie was sixth. As a result I was awarded a bursary to attend the University of Adelaide, then the only university in South Australia, and I could choose my area of study. My special interest in my final years of school was physics, particularly high energy physics. My article in the *Science Journal* was on the cyclotron that had been invented at the University of California, Berkeley by Ernest O Lawrence a few years before. I still clearly remember the small semi-popular book titled *The Cyclotron* that made a great impression on me.[1] However in the event I went into medicine.

I cannot remember whether I ever seriously considered going into physics. Indeed I cannot remember precisely why I chose medicine although we were a medical family with my father a surgeon and my mother a nurse. Possibly there was a feeling there was more economic security in medicine. The fact that I could simply go into medicine because that is what I wanted to do contrasts greatly with present practice in the U.S. where applying for medical school is an enormous hassle. It is true that the more relaxed attitude in Adelaide resulted in some inappropriate students. Certainly one of these was my friend Neil Neilsen whose main interest in life was music. He was an excellent bassoonist and we sat next to each other during the compulsory physics lectures that were a complete waste of time for me because I had studied physics at a higher level during my last 2 years at school. We mainly talked about Wagner and partly because of this I became very knowledgeable about his operas.

I greatly enjoyed the freedom and opportunities of the University. However I did not enjoy the medical course that in many respects was very badly taught. As already mentioned, the first year physics course was a complete farce because

[1] *The Cyclotron*, W. B. Mann, London: Methuen & Co., 1940.

J.B. West, *Breathing on the Roof of the World*, Springer Biographies,
DOI 10.1007/978-1-4939-7122-0_2

Fig. 2.1 Original Mitchell building of the University of Adelaide which dates from 1881. The beautiful campus spreads from North Terrace down to the River Torrens and beyond

attendance at the lectures was compulsory even though I already knew all the material at a higher level. The same was largely true of chemistry where again attendance was required and a roll was taken. Some of the courses were ludicrously taught. For example, in our first lecture on histology, the woman professor started by saying that it was her practice to dictate the notes during the lectures and we therefore began writing and did this for the whole course of 15 weeks or so! I should have enjoyed physiology but did not because it was poorly taught by Stanton Hicks. He was an entertaining lecturer and had some charisma but the subject was taught in a very descriptive way without any quantitative challenge. To someone who had done well in physics and chemistry it lacked any analytical components. This experience set back my interest in physiology by about 10 years. We also took botany and zoology in our first year and these subjects should have been interesting but were not. This was partly because they were taught by faculty who regarded teaching medical students as a chore. None of the students would go any further in these subjects, and in fact most of the medical students regarded the material as irrelevant.

The clinical teaching was worst of all. This took place in the Royal Adelaide Hospital adjacent to the University, and these were the days when there were no full time professors of medicine or surgery. Instead the teaching was done by consultants in private practice (known as "honoraries" because they were unpaid by the university hospital) for whom teaching medical students was a low priority. Some of the teaching methods were deplorable. I can remember sitting through the lectures Wounds 1, Wounds 2, *etc.* on hot summer afternoons after lunch in a poorly ventilated lecture theatre when the only visual aid was an epidiascope throwing a blurred image. I can also remember waiting in the wards for one of these consultants to give us a tutorial, and learning after 40 min or so he was unable to make it on that occasion. In fairness, I should add that this system was about to change, and when my sister, Rosemary[2] did medicine 2 years later, there were full time professors of both medicine and surgery.

Another reason why medical school was disappointing was that the class had two very different kinds of students. There were those of us who had come straight from high school, and of course we were used to working hard if the material was interesting and challenging. However there was a large group of returned service men who had just been demobilized after the war and were taking the medical course. These were all 4 or 5 years older than us (a large difference when you were 18 as I was) and understandably their main interest was to get through the medical course as quickly and painlessly as possible and get on with their lives. The scientific by-ways that the course might lead you into were of no interest to them. This odd mixture of students probably blunted the intellectual curiosity of the class.

Although I disliked the medical course, I took full advantage of other opportunities in the University especially in the first 3 or 4 years. In my first year I enrolled in a philosophy class for credit, and also a class in harmony in the Music Department. Since the medical course was full time I must have missed a few classes. I enjoyed the harmony class although I thought the pace of teaching was extremely slow. However the most important extracurricular activity was my interaction with a group of people through the Student Christian Movement (SCM). This was not an evangelical religious society but more a group of people interested in the social and political issues of the day, usually with a leftwing bias that was very common at the time. It included some wonderful people who have remained my friends until the present day. These included Colin Taylor, Roger Opie, Beryl Martin, Tom and Elizabeth Allen, David and John Dunn, Elizabeth Prest, Dick Rischbieth, Bronwen Murdoch and many others. We used to have occasional weekend meetings at a small conference center in the Adelaide Hills and there was a great sense of fellowship. It was also a very musical group. I was very active in the SCM and was president at one stage, although towards the end of my medical course I drifted away somewhat. Whenever I returned to Adelaide I looked forward to seeing the survivors of this group who were all friends of my cousin Ren Potts.

[2]Rosemary excelled in her medical course and went on to practice as a general practitioner. Later she became a psychiatrist.

Fig. 2.2 Adelaide University hockey team during an interstate carnival in Sydney in 1951. I was captain of the team. On my *left* is Brian Ellis who became Professor of Philosophy at Latrobe University in Melbourne

I was fairly active in university sports and became a strong field hockey player (Fig. 2.2). I eventually captained the university team and we had two hockey carnivals out of state in Melbourne and Sydney. The one in Sydney was particularly memorable because I met a young philosophy student from the school of free thinkers who dominated the Department of Philosophy at the University of Sydney at the time. I also played junior hockey for South Australia at a carnival in Perth and clearly remember the three-day train ride where we went for hours across the Western Australian desert known as the Nullarbor Plain. The railway line is dead straight for about 300 miles! I also had some interest in athletics and ran the mile occasionally.

I became very fond of music during my period at Adelaide University. The music department in a building known as the Conservatorium had a fine collection of the old 78 r.p.m. gramophone records that had been donated by the Carnegie Foundation in the U.S., and I used to go in with friends to play some of these early in the morning before classes began. Later I belonged to a private library that lent gramophone records. The University had a choir that was conducted by a physicist, John Prescott, using a screwdriver as a baton. A high point was rehearsing "And the Glory of the Lord" from Handel's *Messiah*. Adelaide had a strong musical community and we had excellent symphony concerts. Although Australia was very isolated from the rest of the world, the Australian Broadcasting Commission would bring out well-known artists and arrange for them to tour through all the capital cities giving concerts. I was also one of the first members of the board of the Musica Viva Society that organized chamber music concerts. I think they wanted a student member and I was friendly with the daughter of one of the people who was responsible for organizing this group which still exists. As mentioned before the SCM group was very musical and I clearly remember Colin Taylor playing the accompaniment of "Much Binding in the Marsh" which was a much-loved B.B.C. radio program while we belted out a song composed for the occasion. We were greatly influenced by the B.B.C. at that time.

I finished the medical course in December 1951 at the age of 23. Because I disliked the medical course so much, and because I had so many interests outside medicine, I did not do well in the final exams although I swotted quite hard for them. Consequently when the results came out it looked as though I would not be able to get a residency position which was an embarrassment to my father who had a fairly prominent position on the medical faculty. However fortunately someone withdrew and I was accepted into the residency program. I then spent 1952 doing a rotating residency with periods in medicine, surgery, and various other areas including mental diseases. One of my clearest memories was a month spent in a mental hospital in the Adelaide Hills where there was an unforgettable female patient with manic-depressive disease. During her manic periods she was tremendous fun and I remember the superintendent warning me not to be seduced which apparently was very much a hazard in these circumstances. I also spent a month in a tuberculosis sanatorium where I refilled therapeutic pneumothoraxes in patients whose lungs had purposely been collapsed to assist healing! This was a time when antibiotic therapy for tuberculosis was in its early days. Modern medical students find this difficult to believe just as they also cannot believe we had no arterial blood gases in those days. I do not remember much of the medical and surgical rotations although we did have a patient with a bleeding peptic ulcer who very nearly died in spite of enormous amounts of transfused blood. There was always tension between the physicians (internists) and surgeons about the treatment of these patients each believing that the other was misguided. My main memory of the surgical rotation was the extreme boredom of assisting with long operations while I was holding a retractor. However I do remember one operation for some kind of renal disease when the surgeon removed the appendix by error, on autopilot as it were, because he was so busy talking to his assistant. Of course this did not matter—the appendix was better out than in. Between finishing my year of residency and leaving for England I did a couple of locums including one in an Adelaide suburb. This involved one or two unpleasant experiences including an ineffectual attempt to treat an infant with vomiting and diarrhea.

Early on I had made up my mind that as soon as my year of residency was finished and I could therefore be licensed to practice medicine, I would move to London. The Adelaide scene had become increasingly unattractive in large part because it seemed constricting and I wanted to see more of the world. It was very common at that time for Australians to go to the U.K. I knew that I could easily become licensed in England and I had become increasingly anglophile. Those were the days when essentially all the history at school was European, chiefly English; I never learned any Australian history. I had several friends in the U.K. including Colin Taylor, Mary Weaver, and Beryl Martin. As a result I made arrangements to sail to London as soon as this was feasible in 1953. At that time nobody went by air presumably because it was so expensive. I still have my first passport which was issued on January 19, 1953. My 1953 diary, which I also still have, notes that I left Adelaide on February 18 to take the train to Melbourne so that I could board the boat, the Moreton Bay, which sailed on February 20.

In retrospect it is perhaps remarkable that I took off with no job in sight and only a few Adelaide friends in London. However in those days Australians could take up residence in the U.K. freely. The cover of my passport said "British Passport: Commonwealth of Australia" and inside I was described as "Australian citizen by birth and a British subject." As indicated earlier, I also knew that I could be registered as a medical practitioner in the U.K. based on my medical degree and one year of residency in Adelaide. So there was no doubt that I could earn a living.

My parents gave a party for me before I left and one of my memories is my cousin Robert (Bob) West who was a successful physician in Adelaide (he later attended the Governor of the State) advising me to get my membership of the Royal College of Physicians in London. My father also emphasized that. Bob had spent some time at the Postgraduate Medical School, Hammersmith Hospital in London and wrote to one of the prominent physicians there, Richard Bayliss, asking him to help me. This turned out to be my only professional contact in London but it was a critically important one. I think I had the notion that a young man with an honest face was bound to make his way in London, but alas that was a misconception.

Chapter 3
London and Llandough, 1953–1956

I arrived in Southampton on March 30, 1953. The sea voyage therefore took just a month and I was traveling on the cheapest ticket that I could get. The voyage was extremely miserable at first because I had severe seasickness and spent several days in my bunk which was one of four in a small cabin. There seems to be a travel sickness gene on my mother's side—one of her sisters used to get sick in a car on a drive around the block. However later I got my sea legs and it was not long before the voyage became just boring. Nevertheless there were a number of pleasant young people on board and indeed we later kept in touch in London for a while. One of the nicest people was a young American who had been in Australia on some sort of agricultural traveling scholarship. He naturally fell in love with one of the pretty Australian girls. We called at Colombo in Ceylon (now Sri Lanka) and foolishly I did not take the opportunity to visit Kandy in the hills, although at the time I had very little money and I was not sure when I would start earning again. I remember that we also docked at Suez and some people went overland to Port Said while we negotiated the Suez Canal. It was a relief to finally arrive in Southampton.

I took the boat train to London on the morning of March 31 and my first stop was the National Bank of Australasia in the Strand which was adjacent to Australia House. This was a great meeting point for Australians and there were large numbers of these in London. I think it was someone at the bank, or it might have been Colin Taylor, who found a bed-sitting room for me in 33 Courtfield Gardens (Fig. 3.1) not far from Earls Court tube station. That address turned out to be my home, off and on, for a number of years although I gradually moved out of the basement into slightly more salubrious bed-sitting rooms. Colin Taylor had a bed-sitter nearby and we saw a good deal of each other. He had arrived in London several months before and had developed an interesting circle of musical friends including another Australian, Geoffrey Parsons, who later became a very successful accompanist and made many recordings with eminent lieder singers. I remember a party where Geoff played the piano and hit the keyboard with such force that I thought it would end up on the floor.

J.B. West, *Breathing on the Roof of the World*, Springer Biographies,
DOI 10.1007/978-1-4939-7122-0_3

Fig. 3.1 Townhouse at 33 Courtfield Gardens, Earls Court where I had a bed-sitting room for a number of years. Number 33 is in the middle of the row

My diary shows that I enquired about locums on that first day, and on the following, I visited the General Medical Council office to apply for registration. I also applied to the Medical Defense Union for liability insurance. My registration was effected in about a week—one of the enormous advantages of being part of the Commonwealth in those days. However at about the same time I arranged to see Sir Francis Fraser who was head of the British Postgraduate Medical Federation that had been set up to assist doctors who wanted to do postgraduate training in London. The first thing he asked me for was a letter of reference from his contact in Adelaide. Since I did not have one (and in fact had left under something of a cloud) he was about as rude and humiliating as only an Englishman can be to a colonial, and I never forgave him for this. He could easily have been more helpful.

By April 20 I had begun a locum at the South West Hospital near Clapham North tube station. The "locum tenens" system still exists in London and was a boon to new arrivals like myself. The opportunities occurred because junior house staff were entitled to vacations, and the periods usually lasted from 2–4 weeks or even more. Because the locum was there for such a short time, the responsibilities were never great but you were assured of board and lodging and other perks such as free laundry. The South West Hospital which still exists on the A to Z map was pretty rundown in 1953 and was built in the old "Nightingale" style of a very long central corridor with wards branching off it. However it was a relief to be employed and as far as I can recall I enjoyed my time there.

My plans were to spend an extended period on the Continent but eventually I had to find a permanent position. My diary reminds me that on April 28 I went to Hammersmith Hospital to see Dr. Richard Bayliss. As narrated earlier, he was a friend of my cousin Bob West who had written to him about me. Bayliss was an impressive man who later became physician to the Queen. I clearly remember taking the Underground out to Hammersmith station only to find that I was about two miles from Hammersmith Hospital which was not in Hammersmith at all but Shepherd's Bush, and that I was running short of time. I had to spend two or three of my hard earned pounds to take a taxi but eventually had tea with Bayliss in the old Lower Medical Corridor. He was supportive but adamant that an honest face was not enough to get on in London. He made the point that the only medical school where I was likely to get a position for good further training was the Postgraduate Medical School at Hammersmith Hospital which had a tradition of catering to Commonwealth students. However he pointed out that I was far too green to get a house physician appointment with my minimal experience. In fact the situation was so competitive that some house physicians there already had their membership of the Royal College of Physicians (MRCP.)! He said that the best plan for me would be to take another house physician job in a hospital where the consultant had strong links with Hammersmith. If I made a good impression that person could then write a letter of recommendation and I might be able to get a position at Hammersmith. The man he suggested was Harry Kopelman who was a consultant physician at St. Margaret's Hospital in Epping, a few miles northeast of London. He had spent a good deal of time at the Postgraduate Medical School where he was well liked. Bayliss' advice was critically important for me in getting a foot in the door at Hammersmith, and that was critically important for my subsequent career.

Another hospital where I did a locum at this time was the old Western Fever Hospital in Seagrave Road near Earls Court. The hospital has now been demolished. While I was there I bought an old used car, a tiny Austin Seven,[1]

[1]The first Austin Seven was produced in 1921. The small four cylinder engine had a capacity of about 700 cc and developed 7.2 hp. In 1924 the engine capacity was increased to 747 cc and that was probably the one I owned. Many Austin Seven clubs still exist in the U.K. and also in other countries.

Fig. 3.2 The Austin Seven car that took me from London to Venice and back on my Grand Tour of the continent. This photograph was taken in Heidelberg

also known as a Baby Austin (Fig. 3.2). For a long time I had planned an extended visit to the Continent by car and in retrospect it was very optimistic to expect this ancient tiny two-seater to make the long trip. My plan was to leave for the European Continent shortly after the upcoming Coronation of Queen Elizabeth II on June 2. Also at the time I was very keen on the music of Benjamin Britten and tried very hard to get a ticket to his new opera *Gloriana* that was opening at Covent Garden in connection with the coronation but I was unsuccessful.

I set out from London in the car on June 6 with the idea of staying in youth hostels as far as I could. However there was not one at Dover so I had to spend the night in the car. I took the car ferry from Dover to Calais the following day and remember that the sailor who insisted on driving the car on to the ferry was surprised that the brakes were so poor that he ran into the car in front! Frankly the car was only marginally roadworthy.

The trip on the Continent was one of my most memorable adventures. I still have my book of maps *Europa Touring* and am only sorry that I did not mark the route because I am a little vague about some parts of it. From Calais I drove through France to Belgium and remember spending a night in Bruges. Of course this was long before the European Union came into existence so my passport has all the appropriate entry and exit stamps. On June 9 I reached Germany, probably at Aachen, and made my way to Dusseldorf and then to Wuppertal. My objective was to visit Sigrid Falke-Rohen's home near the village of Fredeburg in the Sauerland in Westphalia. She was the German nurse-assistant that I had known in Adelaide, and she invited me to visit her in her home. She had suggested that I ask the way in a bar in Wuppertal but when I did this my destination turned out to be another 50 miles or so east. Nevertheless I eventually arrived at Gut Altenhof, Sigrid's family home on June 11 and stayed there for 2 weeks. This was a wonderful visit. It was a large house with various surrounding buildings in the middle of a wooded area and the Falke-Rohens were obviously well-to-do. The buildings included a smokehouse where the family prepared their dried meats. One of the most memorable evenings was a "schutzenfest," an old German traditional event originally related to a shooting competition. There was prodigious drinking of beer, wine and some kind of schnapps. I remember someone telling me that the people of Westphalia were born two liters of beer under (the implication being that they needed a couple of drinks before they were sociable). However I met many interesting people. I have always enjoyed being in Germany possibly because of genes on my mother's side. Incidentally I wrote to Sigrid 2 or 3 years ago only to find that she was deceased. The Falke-Rohens still live in Gut Altenhof and they invited me to visit if I am in the area.

After 2 weeks in Gut Altenhof I set off traveling south on the autobahn from Cologne, and one of my memories is a spectacular youth hostel in a castle in Koblenz overlooking the confluence of the Moselle and Rhine rivers. A less enjoyable feature of the stay was that a group of Germans in the same room never stopped talking all night. Then on to Heidelberg where Sigrid had given me an introduction to one of her friends. I have vivid memories of a magnificent party at the famous castle in Heidelberg where there was a torchlight procession around the castle ramparts, a good deal of drinking, and the unforgettable sight of people dancing on an enormous wine vat. I also remember walking up the Philosophen Weg (the Philosopher's Way) which is historic.

Further south I had a major problem with the car when the rear axle broke. Parenthetically I doubt if I ever had the car properly serviced before I left and it was fortunate that this was the only serious problem. Happily there was a very enterprising German garage nearby and they were able to make a repair. In retrospect this was pretty remarkable as I am sure they did not have spares for baby Austins. However after 2 or 3 days I was off again and entered Switzerland at Basel. My passport shows an exit stamp for Germany on July 8.

I spent several days in Switzerland seeing the sights and staying in youth hostels all the time. One of my memories is the Lion of Lucerne which I had to see because it was prominent in a children's book which was popular with my father. But of course I was also duly impressed by the magnificent alpine scenery. Eventually I drove over the Simplon Pass into Italy. My little car with its puny engine found the

Alps heavy going and I remember toiling up the approach to the Simplon Pass in first or second gear. Then on to Milan and eventually over to Venice where there was a rather grubby but colorful youth hostel on an island across the water from St Mark's Square. Incidentally when I parked the car in the huge garage at the entrance to Venice, I noticed that a couple of the tires were worn out. It was extremely fortunate that I was able to find second-hand tires of the appropriate size there.

By now it was mid July and very hot but I duly visited many of the sights. During the course of the trip I made friends at a number of the youth hostels but never had a traveling companion. Some of the hostels had strict rules about arriving on foot so I used to park my car a short distance away and pretend I was hiking. I think I probably had plans to go to Rome but was beaten back by the heat.

From Venice I turned north again and made my way back over the Alps to Munich where I had a good friend, Adrian Cook, from Adelaide University. He was very musical and was studying German and music in Munich. I am not sure how I first met him at the University but I saw quite a bit of him in my final year as a medical student because he was a patient in one of the TB wards with pulmonary tuberculosis. He must have made a good recovery although he was always a rather pale, sensitive, artistic person. He arranged some accommodation for me with a friend in Munich and we had three or four very enjoyable days together there. I vaguely remember a party at Schwabing in the artist's quarter. Adrian had a colorful circle of friends, mainly musical.

We then set off together by train to attend the Wagner festival in Bayreuth. These were early post-war days for the Bayreuth festivals and I did not realize at the time how lucky I was. I cannot remember exactly how I arranged the tickets (these are difficult to come by) but it may have been through Adrian. We saw the *Ring* cycle and I also heard *Parsifal*. As mentioned earlier I had become very fond of Wagner and knew the operas intimately including the music and the librettos. Bayreuth was an unforgettable experience. The days were completely taken up with Wagner. We talked about the operas in the morning, the performances went from mid-afternoon until some time in the evening with long breaks between the acts for meals, and then we would repair to the café Die Eule for coffee afterwards. This is said to date back to 1444. I remember seeing some of the singers there with great wraps around their necks to protect their voices. A special treat was being able to attend one of the rehearsals which I still clearly remember. This was normally verboten but Adrian knew one of the string players and we were able to sneak in with the orchestra.

I stayed in a private house in Bayreuth and the woman who owned it, Frau Fricke, offered to draw my portrait using crayon. This she did over several sittings and I still have it (Fig. 3.3). It is excellent. However there was an unfortunate incident in the middle of one of the operas when I was sitting next to Adrian. He said he felt ill and had to leave. This caused something of a commotion because the opera house did not have a central aisle and he had to negotiate many sets of legs. When I saw him after the opera was over he had a rash which I seem to remember was chickenpox. I put him on the train back to Munich where he arrived safely enough and made a good recovery although there was some criticism when I returned that I had not traveled

Fig. 3.3 Crayon drawing made while I was staying in the house of the artist in Bayreuth

back with him. Bayreuth was certainly the climax of my Grand Tour of the Continent and shortly after I returned to Munich I set off to drive back to London which I reached on August 16. I had been abroad for 10 wonderful weeks.

By this time my coffers were low but I was able to find a locum almost immediately and started at a hospital in southeast London 2 days after I arrived back. This was followed by another 2 weeks at May Day hospital in Croydon, and this was the only locum that left a sour taste in my mouth. Generally the locums were fun because I met many doctors of about my age and of course as a temporary physician you had few responsibilities. However for some reason the atmosphere of the May Day hospital was unpleasant with various personal animosities. I also had a nasty experience while giving an anesthetic with unfamiliar equipment. That was the last locum I ever did.

As narrated earlier, my plan was to spend a period as house physician at St Margaret's Hospital in Epping with Harry Kopelman on the recommendation of Richard Bayliss. On arriving back in London I saw an advertisement for this position and drove out to Epping where I lost my way but arrived at the interview in the nick of time and got the job. I started this 6-month appointment on September 16, 1953.

The arrangements for this house physician job were unusual. The first 3 months were spent at Honey Lane Hospital near Waltham Cross, a small facility for non-acute cases taking the overflow from St. Margaret's Hospital. These were typically patients with peptic ulcers or perhaps they were recovering from heart failure. In those days there was no great urgency to discharge people from hospital. Prior to the National Health Service, this small hospital had had its own medical superintendent with a splendid two story house in which I now lived. There was a maid who came in, cleaned, and served my meals and I had a roomy garage for my car. It was a very elegant setting indeed for someone who had just spent a couple of months in youth hostels. Every morning Harry Kopelman or another consultant from St. Margaret's Hospital would arrive and we would do leisurely ward rounds. I was the only doctor in the hospital and was therefore treated royally by the nursing staff. When I drove to St. Margaret's Hospital which was a few miles away through

Epping Forest two or three times a week, I would occasionally come across deer crossing the snowy road.

After 3 months at Honey Lane I moved to the main hospital on the edge of the town of Epping and there the pace increased dramatically. There were several house physicians and the occasional registrar, and a good deal of acute and semi-acute medicine. I very much enjoyed working with Harry Kopelman who was a good physician with a strong interest in cardiology. Several of the resident physicians were colorful including the chief resident surgeon who was anxious to increase his experience with gastrectomies (removal of the stomach), and I fear that two or three of the local citizens may have lost their stomachs unnecessarily. There were several other interesting resident physicians at St. Margaret's Hospital although I have now lost track of them. One was a very elegant and highly intelligent black woman physician from Africa who must have gone on to a very bright future.

My 6 month stint at Honey Lane and St. Margaret's hospitals finished in March 1954 and by then I think I had decided that I must get my membership of the Royal College of Physicians if I was to have any sort of career in academic medicine. This is equivalent to internal medicine boards in the U.S. I decided to take advantage of the courses run by the British Postgraduate Medical Federation and attended two courses, one at the National Heart Hospital in Westmoreland Street, and another at the National Hospital for Nervous Diseases in Queen Square. Both of these courses were extremely well done. At the Heart Hospital I attended outpatient sessions run by Paul Wood who was one of the top cardiology diagnosticians in the country and probably in the world. It was fascinating to see him evoke physical signs and discuss the importance of determining whether, for example, a cardiac murmur was present or not before going on the next sign. Wood was of English parentage but trained in medicine in Melbourne and had a blunt somewhat confrontational Australian manner. Although he was a brilliant clinician he made many enemies. I remember on one occasion when a case was being presented to him, one of the registrars was unable to produce an X-ray or whatever and he received the most withering criticism from Wood. I thought the best thing that registrar could do then would be to emigrate! Later on I saw Paul Wood demonstrate his talents at Sully Hospital in South Wales where there was a cardiac surgery unit. My cousin Len West (brother of Bob West referred to earlier) was one of the consultant physicians there and I visited the hospital on several occasions. Their practice was to collect all the most difficult diagnostic problems over 2 or 3 months and then invite Wood down for a day. The patients were lined up on one side of the room and Paul Wood would go from one patient to another. He was given only a brief history and shown the electrocardiogram and chest X-ray (essentially the only ancillary data available at that time), and he would then do a brief physical examination and announce the diagnosis. This was an extremely challenging test in a situation where he had plenty of enemies and it was a magnificent performance.

The sessions at Queen Square were equally entertaining and impressive. The top British neurology consultants were there and they would bring in patients with unusual conditions, elicit the physical signs, and then discuss the diagnosis. Some of the patients had very unusual conditions and a few cynics in the audience thought that some of them might be professional actors! I wish I could remember the names of some of the eminent neurologists.

At that time I lived in bed and breakfast digs in Bloomsbury. Having read novels by the Bloomsbury group I had a very romantic notion of the area but of course it had completely changed by 1954 and the digs could hardly have been more miserable. It was at this time that I was invited to a wild nurses' party at the Central Middlesex Hospital just outside London. The goings on would not be appropriate for this account but I did meet a very nice nurse, and we remained friendly for a year or so.

I then enrolled in a postgraduate course given by the Postgraduate Medical School (P.G.M.S.) at Hammersmith Hospital (Fig. 3.4) in the hope that, having taken the course, and with a letter of recommendation from Harry Kopelman, I would be able to obtain a house physician position there. These courses were attended by many young doctors from overseas including a number of Australians, and my friend David Dunn from Adelaide had also enrolled. David and his wife Ann invited me to stay with them as a paying guest in their rented flat on St. Quintin Avenue just east of Wormwood Scrubs. This was very convenient to Hammersmith Hospital but in one of the dingiest parts of London. Ann was essentially completely deaf but was able to communicate with David by lip reading and some sign language. She was the daughter of Dr. Southward (Happy Jack) who was a consultant physician at the Royal Adelaide Hospital, and she had been working as a student in one of the bacteriology laboratories when she developed tuberculous meningitis. This would have been fatal were it not for the fact that the antibiotic streptomycin had just been introduced. This cured her T.B. but unfortunately left her deaf. David and Ann were wonderful hosts and we had many good times. A few years ago she was fitted with a bionic ear which apparently worked well. I did not give the course at the P.G.M.S. very high marks but at the end of it I applied for a position as house physician and was accepted. Naturally I was elated. I stayed with the Dunns' until I was able to move into the hospital.

The Postgraduate Medical School had an impressive reputation. It had been started in the thirties by Sir Francis Fraser and soon acquired a number of absolutely first class academic physicians. From the beginning, all the faculty had full time academic appointments, that is they did not have the right of private practice. I think that the P.G.M.S. may have been unique in that respect in London and perhaps the U.K. Certainly all the big London teaching hospitals were staffed by physicians with private practices, although there were two or three exceptions in the professorial units. The P.G.M.S. developed a tradition of strong clinical excellence with first-rate research and was certainly in the vanguard of academic medicine in the U.K. However among the London teaching hospitals it was considered something of an upstart and was controversial. This was partly because of the colorful characters on the staff (faculty) that included, for example, Sharpey Schafer who was said to favor odd socks of different colors. But the standard of clinical presentations at the Staff Rounds as they were called (grand rounds) was extremely high with excellent uninhibited discussion. Among the luminaries were John McMichael who later became F.R.S. as did John Dacie in hematology. Other

Fig. 3.4 Hammersmith Hospital in DuCane Road, Shepherd's Bush. I spent 15 years there, off and on, from 1954 to 1969

well-known names included Sheila Sherlock, Russell Fraser, Malcolm Milne, Charles Fletcher, Eric Bywaters, John Goodwin, and Cuthbert Cope.

Just as the consultant physicians were eminent, the registrars, and even the house physicians were impressive. One of the senior registrars and good friend was Christopher Booth who subsequently had a distinguished academic career. One of my best friends on the house staff was Jim Lance who later became a well-known consultant neurologist in Sydney. The house physicians at Hammersmith included a broad mix of people from the Commonwealth because this was the best place in London to learn academic medicine. Indeed this was one of the reasons for setting up the British Postgraduate Medical Federation of which the Postgraduate Medical School was a part.

During this time, I first met Tim Gordon who was in the Pathology Department. Tim had a flat in an elegant row of Georgian houses in Hammersmith Terrace overlooking the Thames, a very attractive area. This became something of a center for people from Hammersmith and it was Tim who later found a flat for me to rent in Kelmscott House in Chiswick Mall just down the river from Hammersmith Terrace.

I was the house physician for Dr. Cuthbert Cope whose main research interests were related to steroids. I did not enjoy the clinical side very much. I have never been a good clinician and found the routine physical examinations tedious. However I did well enough. Near the end of my period with Cope I asked him what he thought I should do. My main strengths were in physics including mechanics, that is pressures, flows, resistances, elasticity and the like, and so it was natural for me to go into either cardiology or perhaps respiratory medicine. Cope said that I should talk to Charles Fletcher whose beds were on the other side of wards B5 and B6 where Cope's beds were. So I went to see Fletcher and he told me that the P.G.M.S. was planning to embark on a new program of research in respiratory medicine. Another Australian, Kemp Fowler, was building a mass spectrometer specifically for respiratory research, and Philip Hugh-Jones who had worked on respiratory physiology at the Pneumoconiosis Research Unit (P.R.U.) in South Wales would be coming to Hammersmith to set up the new program with support from the Medical Research Council. Fletcher suggested that I spend a period at the P.R.U. learning respiratory physiology and then return to Hammersmith. Fletcher himself had been the first director of the P.R.U. a few years before, and of course knew everybody there and was able to give me a strong recommendation. In the event this is the course I took, and it was another in those critical decisions that formed my career.

In the spring of 1955 I drove down to the Pneumoconiosis Research Unit that was based in Llandough a few miles west of Cardiff. Before I went there various people remarked how much I would enjoy meeting the Welsh but in fact the P.R.U. was an English enclave with almost no Welsh people at all. It had been set up by Charles Fletcher as a Medical Research Council Unit to investigate the lung disease that occurred in coal miners because of the dust, so-called coal workers' pneumoconiosis. This is also known as black lung disease in the U.S. There was a long history of coal mining in South Wales, and the Welsh mining villages were notorious as a source of pneumoconiosis. The villages tended to be isolated because only one road came into the valley and this was also the only way out. The work underground was hard and demeaning, but such was the tradition of the villages that for many of the boys their main ambition was to follow their fathers into the mine. Lung disease among the coal miners was rampant, and there was also a lot of pulmonary tuberculosis.

My year at the P.R.U. was one of the most enjoyable in my life. At that time the unit had an excellent scientific staff headed by John Gilson who had taken over from Charles Fletcher. Notables included Archie Cochrane whose name is now a byword in epidemiology, Martin Wright, nominally a pathologist but really an inventor whose name is perpetuated in the Wright respirometer, the statistician Peter Oldham, and several well-known physiologists including John Cotes, Colin McKerrow and, for a period, Bill Briscoe. Some of these people had been at the Institute of Aviation Medicine at Farnborough during World War II where respiratory physiology flourished. John Cotes had helped to design the oxygen masks for the first successful ascent of Everest in 1953.

Several of the scientific staff were colorful. For example, Archie Cochrane had been a member of the International Brigade in the Spanish Civil War! One of the stories was that he had parked his car across the road leading from London airport to obstruct Neville Chamberlain when he returned from Munich with his "piece of paper" that ostensibly proved that Hitler did not have warlike intentions. Archie was unmarried and lived in a splendid farmhouse with a manservant. Peter Oldham was very jolly and frequently took snuff from an elegant box after which he blew his nose violently into a red silk handkerchief that he kept in his sleeve. Another memorable person was Vernon Timbrel whose primary interest in life was the properties of coal dust. He was one of the world's experts and had enormous enthusiasm for his subject—a good advertisement for the fact that any scientific topic becomes interesting if you delve deeply enough into it. One of the most eccentric people was Bill Briscoe who became a good friend of mine. He had previously been at Hammersmith and had also worked with Andre Cournand (Nobel Laureate for his work on cardiac catheterization) in Bellevue Hospital in New York City. Bill had all sorts of strange customs such as suspending his pencils from the ceiling above his desk with rubber strips so that they were instantly available. He would also send postcards to himself as reminders of various things. Bill subsequently moved to New York where I visited him in 1958. Unfortunately he was a heavy smoker and died of cancer of the larynx. I worked mainly with John Cotes but the project was not an inspiring one. I measured cardiac output in the coal miners using Evans blue dye although the scientific justification for this eludes me now. One of the byproducts of this project was that the miners whom we studied developed a blue discoloration of their skin. Nothing publishable came out of this but I did learn a great deal of respiratory physiology.

The M.R.C. Unit was very well organized. It had its own building attached to Llandough Hospital and was set up purely for research. I had never been in a dedicated research environment before and found it enormously stimulating. There were all sorts of helpful features. For example, you simply gave the librarian a list of journals that you would like to see and every issue was delivered to your "in" basket in due course. There were excellent seminars and very good technical assistance. One of the interesting groups was that responsible for producing top class chest radiographs. The diagnosis and staging of coal workers' pneumoconiosis depended almost entirely on the chest X-ray and great efforts were devoted to obtaining the best possible films. Outsiders like myself were not welcome in this suite in case we contaminated the developing tanks or whatever.

A very important event for me occurred in August 1955 when Julius Comroe's book *The Lung* first appeared. I still have the copy that I acquired at that time and the book had an enormous influence on me. It persuaded me to continue in the area of respiratory physiology, and this was a good decision because the subject made enormous advances in the 30 years or so from that time. Comroe later brought out a second edition but in my opinion this was not nearly as good as the first because he tried to make it more clinically relevant and this was not his forte. Incidentally Comroe's book was the standard text in respiratory physiology when I joined the UCSD Medical School in 1969, and because I thought it could be improved upon I wrote my own book *Respiratory Physiology: The Essentials*, the first edition of which came out in 1974. This then took over as the most popular textbook in the area for first year medical students and has remained so.

Although the bulk of my time was spent learning respiratory physiology and doing research, I also had clinical responsibilities in the ward of the hospital that was used by the P.R.U. This was filled with coal miners many of whom had advanced coal workers' pneumoconiosis, so-called progressive massive fibrosis. I think my title was senior house officer (S.H.O.), which was one step above the lowliest title of house physician. Note that I had already done three house physician jobs, one in Adelaide, a second in Epping, and a third at Hammersmith! There was another S.H.O. called John Gorringe and we became good friends. We often formed a foursome with two Welsh girls and went to informal dinner dances.

One of the memorable events that occurred in the hospital ward was the singing of the miners on Sunday evenings. They were all members of the local chapel of their village and the singing was glorious in the Methodist tradition. They sang in parts with great resonance. I also remember going to a rugby match at Cardiff Arms Park where again the singing before the game was very moving.

The Welsh mining villages at that time still had a low standard of living. Archie Cochrane carried out an epidemiological survey looking for pulmonary tuberculosis in a mining village called the Rhondda Fach. This involved taking diagnostic chest X-rays of all the inhabitants. As he related, X-raying the first 50% or so was easy, the next 20% more difficult, the next 10% very challenging and so on until he got to the last person who was the grandmother in an attic with open pulmonary tuberculosis who had been infecting all the children! These surveys were some of the earliest in epidemiology, and I can remember seeing the punched cards running through the Hollerith machine so that those with a particular punched hole dropped into a bin at the bottom. This was long before computers.

Chapter 4
Postgraduate Medical School, 1956–1960

I returned to Hammersmith Hospital in the spring of 1956 to join a Medical Research Council (M.R.C.) group headed by Philip Hugh-Jones. Philip had been at the Pneumoconiosis Research Unit with John Gilson where they had written a very comprehensive report on pulmonary function in coal workers pneumoconiosis in an official publication known as an M.R.C. *Green Book*. This contained a great deal of new and interesting respiratory physiology,[1] but unfortunately because it was an M.R.C. report it remained essentially unknown to most physiologists around the world. After his period at the P.R.U., Philip had gone to the medical school at Kingston, Jamaica for a period, and he then returned to join Charles Fletcher at Hammersmith. One of my vivid memories on returning was that there was little space to work in or even for me to sit down in—a striking contrast to the palatial facilities of the P.R.U. This Spartan situation was typical of Hammersmith in those days but certainly did not impair the productivity of the institution. At a later date, Philip commandeered the house that had belonged to the hospital engineer, and that had since been condemned as too dangerous to use. We then shared a room on the second storey together with a secretary, and the laboratory in which we used radioactive oxygen was downstairs. However we also had some space in the Lower Medical Corridor and I used part of one of the benches for my "office." Parenthetically it was common for productive institutions to be very crowded. I remember visiting someone at the Massachusetts General Hospital on one occasion and finding to my surprise that his office was an abandoned men's room from which the china fixtures had been removed.

This period from the spring of 1956 to the summer of 1960 was one of the most stimulating and productive of my life. As mentioned earlier, there was a plan to develop a new research group in academic respiratory medicine but I am not clear exactly how this happened. It is true that John McMichael who was chairman of the

---

[1]Gilman, J.C. and Hugh-Jones, P., *Lung Function in Coalworkers' Pneumoconiosis*, London: Medical Research Council, Special Report Series, No. 290, 1955.

J.B. West, *Breathing on the Roof of the World*, Springer Biographies,
DOI 10.1007/978-1-4939-7122-0_4

department of medicine and a very powerful man, had an early interest in respiratory physiology. For example he had developed the helium dilution method for measuring lung volumes when he was in Edinburgh. However he made his name in early cardiac catheterization being one of the first people in the U.K. to introduce this procedure after its initial development in New York by Cournand and Dickinson Richards. John McMichael encountered a great deal of opposition from cardiologists in other London teaching hospitals. In fact Hammersmith developed a reputation for invasive experimental medicine that was strongly criticized by some people. One of the anecdotes was that some of the citizens of Hammersmith wore wrist bands stating, "If I become ill, please do not send me to Hammersmith Hospital." Some years later Pappworth wrote a book strongly criticizing some of the investigative work at Hammersmith and elsewhere.[2]

Part of the plan to develop academic respiratory medicine at Hammersmith was to bring Kemp Fowler over from Australia to build the first mass spectrometer specifically designed for respiratory research (Fig. 4.1). Kemp was a colorful character whom I got to know well. He was a talented design engineer and was successful in producing an impressive instrument that allowed rapid analysis of all the important respiratory gases. Philip Hugh-Jones and I had full control of the experiments to be done with this new instrument and it was a tremendously exciting time. In fact as Philip became more and more involved with clinical responsibilities I was essentially in charge of the application of this new technology. I have always been something of a frustrated engineer and enjoyed working with the mass spectrometer very much. From time to time problems with the vacuum system or whatever would occur and I spent many evenings taking the machine apart and putting it together again.

One of our first projects was to determine how much information could be obtained from simple analysis of expired respiratory gas. Without going into too much technical detail, I realized that there was a close relationship between the respiratory exchange ratio (R) at any point in expiration and the ventilation-perfusion ratio of the lung units that produced this gas. This enabled us to obtain a measure of ventilation-perfusion inequality from the change in expired R. In fact this was the topic of my first publication in 1957 in the British journal *Clinical Science*. It is remarkable that we used the same technique in 2003 to look at ventilation-perfusion inequality in the astronauts in the International Space Station. This work on pulmonary gas exchange introduced me to the wonderful world of ventilation-perfusion inequality and I still have my copy of the book by Rahn and Fenn, *A Graphical Analysis of Pulmonary Gas Exchange*, which is dated May 1956. This book made an enormous impression on my and I fell in love with the topic. Because it was so fascinating but so difficult to understand, I later decided to write a book about it, and this was my first book, *Ventilation/Blood Flow and Gas*

[2]Pappworth, M.H., *Human Guinea Pigs: Experimentation on Man*, London: Routledge & K. Paul, 1967

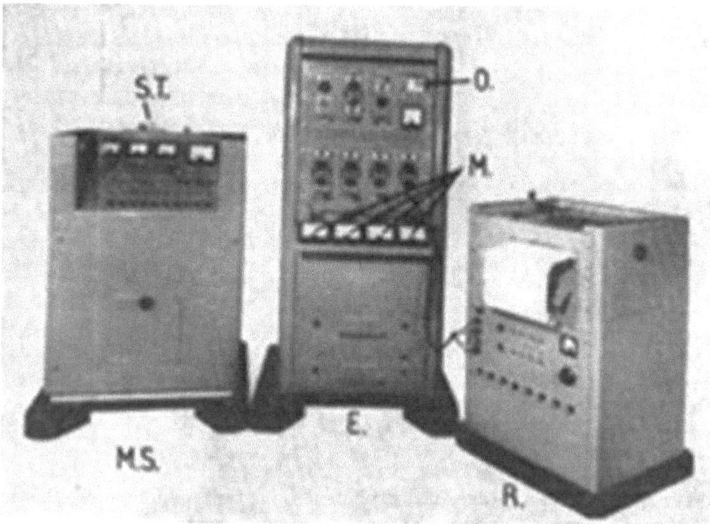

Fig. 4.1 First mass spectrometer specifically designed for respiratory studies. It was built by Kemp Fowler at the Postgraduate Medical School and I was one of the first users. MS, mass spectrometer unit; ST, inlet for the sampling tube; E, electronic cabinet; O, oscilloscope displaying the mass spectrum; M, monitoring meters for the various gases; R, pen recorder

Exchange, published by Blackwell in 1965. It went through several editions and I still get the occasional royalty check.

Shortly after Kemp Fowler got the mass spectrometer into working order, the Duke of Edinburgh (Prince Philip) made a visit to the Postgraduate Medical School and the Dean brought him to our laboratory to see this new instrument. The Duke agreed to breathe over the sampling tube, Kemp was running the chart recorder, while Philip Hugh-Jones was in charge of proceedings. Kemp called out "O.K. Philip let it go" meaning that Philip Hugh-Jones should start the procedure, but the response was that the Duke let out a long expiration that was duly recorded! This misunderstanding caused the Dean a great deal of embarrassment, but the result was a unique recording of royal alveolar gas (Fig. 4.2).

Another development came about at this time because of Philip Hugh-Jones' interest in using bronchoscopy to obtain information about the function of different lobes of the lung. The bronchoscopies were carried out in the Lower Medical Corridor where the mass spectrometer was located, and the sampling tube was passed down through the old rigid bronchoscope into different regions of the lung. This turned out to be a very interesting area of research although it never really produced anything of lasting importance. One of the discoveries was the specific patterns of alveolar gas concentrations caused by airway or blood flow obstruction. Another fallout for me was the invention of a bronchial flow meter that enabled us to measure air flow in different lung regions (Fig. 4.3). The principle was that a small jet of air containing an inert gas such as argon was directed across the bronchial air stream towards the sampling tube of the mass spectrometer. When

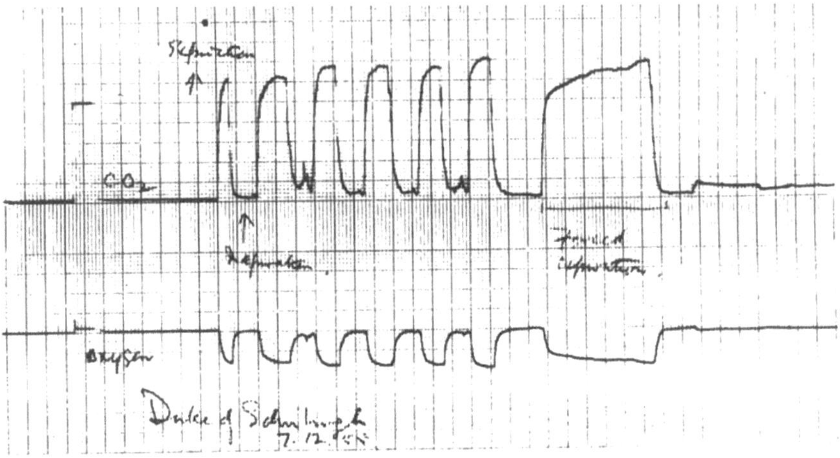

Fig. 4.2 Mass spectrometer recording of an expiration by the Duke of Edinburgh. This may be the first recording of royal expired gas!

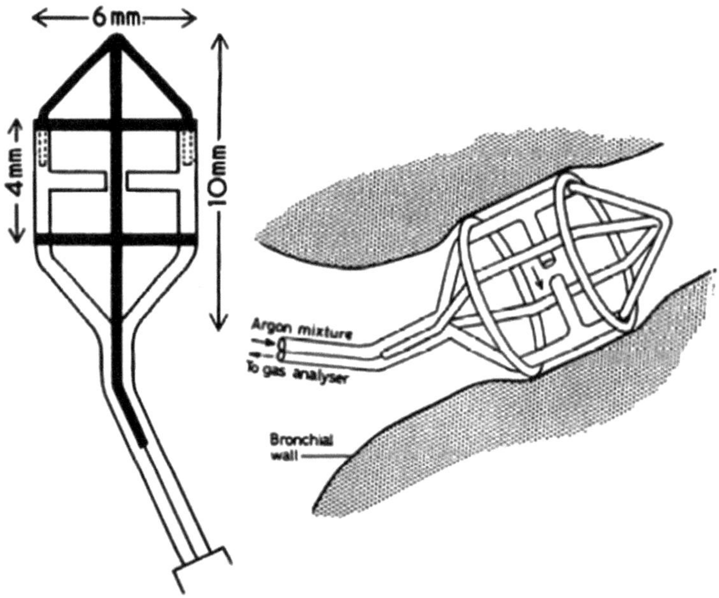

Fig. 4.3 Bronchial flow meter which I invented and patented. Air containing argon is blown across the gap and when this is deflected by the gas flow in the bronchus, the argon concentration recorded by the mass spectrometer is a measure of gas flow

the jet was deflected by the gas flow, the signal from the argon jet was reduced. I remember going into the City of London to get this patented, and being presented with a parchment document tied up with a pink ribbon. There was subsequently some interest by people looking at the exhaust gases of rockets but I do not know if anyone actually used it. The studies in patients with lung disease were compared with measurements on anaesthetized dogs carried out in the same room but at a different time. Philip and I used to drive down to various animal suppliers in south London, bring the animals back in the back seat of the car, anaesthetize them, and carry out the measurements. This was a very amateurish procedure compared with present day practices.

The research with the mass spectrometer was tremendously stimulating and resulted in a number of publications. I also submitted the work as a thesis for my M.D. degree from the University of Adelaide. However at about this time another extraordinary event occurred and this had even more influence on my subsequent career. Some years before, the Medical Research Council had decided to build and install a cyclotron on the Hammersmith Hospital site specifically for medical research. The Medical Research Council Cyclotron Unit as it was called was headed by Derek Vonberg, an engineer, and the group successfully built a 60 inch cyclotron which was situated on the western boundary of the hospital site (Fig. 4.4). The medical justification was twofold. One was to test the therapeutic value of neutron radiation in the treatment of cancer. But equally important was the preparation of short-lived radioisotopes that were readily available by bombarding various elements using the cyclotron beam.

One of the most interesting of these radioisotopes was oxygen-15. This has a half-life of only two minutes (that is, half of it is lost by radioactive decay in two minutes) which means that relatively large amounts can be administered to a patient

Fig. 4.4 The Medical Research Council cyclotron at Hammersmith in June 1955. I believe that this was the first cyclotron built specifically for medical research

or a normal volunteer, but the radiation dose is small because the material decays so quickly. I can still remember a meeting chaired, I think, by John McMichael where it was announced that the cyclotron could now prepare oxygen-15 and, if anyone wanted to use it, there is was!

There is an anecdote about the first production of oxygen-15 at Hammersmith, and certainly parts of it are true. Michel Ter-Pogossian, a physicist at Washington University in St. Louis where another cyclotron had been set up, visited the new machine at Hammersmith and arranged for a bag of air to be placed in the deuteron beam. After a few minutes the bag was removed, he took a breath of the gas, and then placed his hand over a Geiger counter which immediately started to tick. The explanation was that some of the naturally occurring nitrogen-14 in the air had been converted into oxygen-15 by bombardment with the deuterons, and when he inhaled this, it had combined with hemoglobin in the blood and thus reached his hand.

I think the meeting chaired by John McMichael to alert people to the availability of oxygen-15 was probably in early 1957. At any event Philip Hugh-Jones and I lost no time in starting to use it to see whether we could measure regional lung function in patients with lung disease. Initially these were patients with carcinoma of the lung who were undergoing therapeutic radiation. We were able to show that if the subject took a breath of radioactive gas and held his breath, the initial increase of counting rate of counters placed over the chest was a measure of the ventilation of that part of the lung, and the rate at which the radioactivity was removed was a measure of the local blood flow. For these initial measurements we had collimated counters arranged over the front and back of both lungs (Fig. 4.5). Our first paper was probably submitted in 1957 and it was published in 1958. In addition to Philip and myself, the other authors on the paper were Norman Dyson and George Newbery who were both physicists in the Cyclotron Unit. This paper was mainly concerned with the techniques of preparing the radioactive oxygen, detecting the radioactivity including using coincidence counting (a special technique applicable to positron emitters such as oxygen-15), and demonstrating some typical patterns in patients with localized lung disease.

However it soon became clear that the radioactive gas tended to be removed more rapidly from the lower parts of the lung than the upper. To investigate this further we arranged for a series of measurements on normal volunteers including, of course, ourselves. The radioactive dose had been calculated as acceptable by the physicists in the Cyclotron Unit although I must add that this was long before Institutional Review Boards. We found that we could convert the radioactive oxygen into radioactive carbon dioxide with the advantage that the rate of removal of the gas or "clearance rate" as we called it was much increased. The results of these studies were published in a landmark paper by myself and Colin Dollery in the *Journal of Applied Physiology* in 1960. This was the first clear demonstration of the regional differences of blood flow in the lung caused by gravity. It was astonishing to find that the uppermost part of the lung had almost no blood flow in many subjects. Colin Dollery was someone of about my age, also working in the

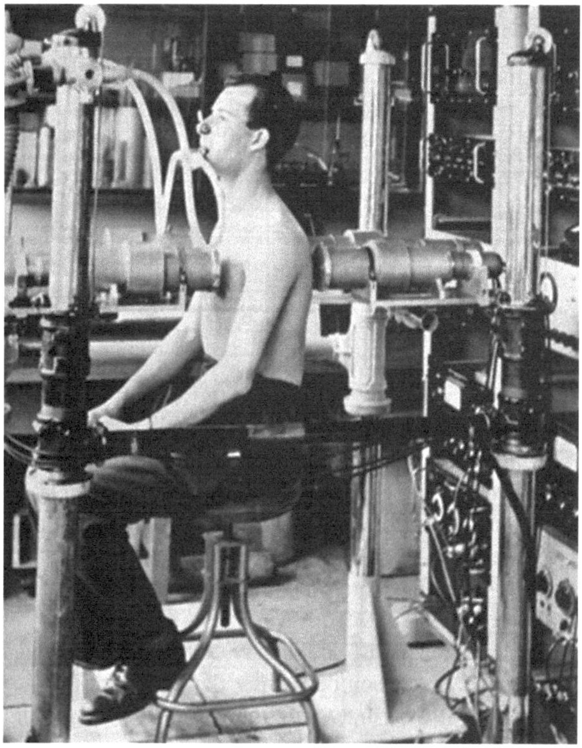

Fig. 4.5 First equipment used to measure the behavior of inhaled radioactive oxygen-15. Scintillation counters were arranged in front of and behind the chest

Department of Medicine, who saw the potential of this work and he collaborated extensively with us.

The demonstration of the striking regional differences of blood flow in the lung, which we showed were caused by gravity, made an enormous impression on both the research and clinical communities. I followed this up by using additional measurements of regional differences of ventilation to derive the topographical inequality of pulmonary gas exchange. We also converted the oxygen-15 into radioactive water vapor and showed that when a breath of this was inhaled, all of the radioactivity was trapped in the upper airways and none reached the lungs. This finding was published in a Letter to the prestigious journal *Nature*. We had another Letter in the same journal on the metabolism of oxygen-15. Incidentally both Colin and myself probably received more radiation than we should have done while carrying out the chemistry to produce these other forms of oxygen-15. However this work with cyclotron-produced radioactive gases established my reputation and I wrote a thesis on this work for a Ph.D. degree from the University of London.

Although this research was by far the most important interest of my life during these 4 years, I made many friends in London and developed an enjoyable lifestyle. I became very friendly with Beryl Dunn who was a physiotherapist at St. Bartholomew's Hospital, but subsequently became the chief physiotherapist to the Royal Ballet. Her very interesting circle of friends included the dancer Margot

Fonteyn. Oddly enough Beryl's father was a well-known orthopedic surgeon like mine. Beryl shared a nice flat in Dilke Street in Chelsea with some other girls including one who was an excellent pianist. We all went skiing in Austria with Inghams on at least one occasion. Through other friends I joined the Holland Park Lawn Tennis Club where I played matches for the Club and I had a wide circle of friends.

After my period as a house physician when I lived in Hammersmith Hospital I moved back to Courtfield Gardens for a period. However Tim Gordon whom I referred to earlier, was living in 9 Hammersmith Terrace on the river, and he introduced me to a Mrs. Stephenson who owned Kelmscott House in Upper Mall which used to be the home of the Victorian artist, William Morris.[3] This had a flat attached to it with a wonderful view of the Thames just upstream of Hammersmith Bridge, and was close to the well-known pub The Dove. The flat consisted of a long reception room with a fine window facing the river (Fig. 4.6) and a bedroom, small kitchen and bathroom at the back. There was a garage below the flat which I had use of. Initially I kept the car there but later I built a 13 foot sailing dinghy in the Enterprise class from a kit (Fig. 4.7). I used to sail this on the river after launching it from the old St. Paul's School floating platform only a stone's throw from the flat. One problem was that you could only sail when the water was near high tide and if the tide turned and you were not quick enough you could be swept under Hammersmith bridge. This happened to me once and there was a very unpleasant noise as the mast bumped along the underside of the bridge and the boat almost capsized. I also became a member of the nearby London Corinthian Sailing Club and occasionally raced on the river.

The flat was very elegant and I had two or three parties there at the times of the Oxford-Cambridge boat races. Of course it was ideal for entertaining with its attractive site and view. The only problem was that in the winter it was cold and damp in spite of continuously running a kerosene heater. I kept the flat even when I was on the 1960–1961 expedition when it was sublet, but this created various problems and when I went to the U.S. in 1961–1962 I reluctantly had to let it go.

These 4 years were one of my happiest periods in London. The research was enormously stimulating, I had a wide circle of friends, my hobbies included tennis, sailing and skiing, and I made the most of the music that London had to offer. This included buying a clavichord although I never learned to play it very well. I also sang in the Bach Choir for several years, which was an unforgettable experience. Charles Fletcher was on the board of the Choir and he introduced me to the conductor, Reginald Jacques. We used to rehearse every week in a hall attached to Westminster Cathedral and I sang many of the great classics including the Bach

[3]William Morris (1834–1896) was a poet, designer, craftsman, and pioneer Socialist, and one of the great innovators of the nineteenth century. He lived in Kelmscott House in Hammersmith from 1878 until his death. He founded the Hammersmith Socialist Society which held weekly meetings in the coach house adjoining Kelmscott House. This was the garage when I lived there. This is now the office of the William Morris Society and they hold meetings there. My flat was above the garage.

Fig. 4.6 My flat in London over the garage of Kelmscott House where the famous artist, William Morris once lived

B Minor Mass, the *St. Matthew and St. John Passions*, the Beethoven *Missa Solemnis* and several other major works. The concerts were initially in the Royal Albert Hall and the echo there was so strong that it didn't matter too much when you came in on your part. However we subsequently moved to the Royal Festival Hall which has a much more clinical sound with no echo, and if you came in a beat early there, the best plan was to emigrate! I enormously enjoyed singing in the Bach Choir and it was a great disappointment when David Willcocks took over as conductor, I was re-auditioned, and immediately ejected! I clearly remember the trauma of the audition. I was asked to sing an unknown piece by sight which was pretty challenging although I thought I did a reasonable job of the melody. However

Fig. 4.7 View from the window of my flat of dinghies racing on the Thames River

Willcocks then asked me to sing the words at the same time which was quite beyond me. It had never occurred to me that anyone could sing both the music and the words from sight. Of course the quality of the choir greatly improved under Willcocks and it made many recordings and traveled all over the world. I can read music fairly well and have a good sense of pitch but their musical skills were way above mine.

In 1958, 5 years after I had left Australia, my father sent me enough money to buy an air ticket to Australia and back. By this time I was very much aware of the important physiology that was being done in the U.S. and I decided to go to Australia via America and come back the same way. This would give me an opportunity of visiting a number of physiologists there. It was a wonderful trip. I flew to New York and having almost no money decided to stay in the Y.M.C.A. that was somewhere in mid-town. The airport bus must have given me a complimentary ride to the West Side Terminal which existed in those days. However I then had to get a bus across town to the Y. I vividly remember getting on the bus with my largish suitcase and giving the driver a $10 note. He did not say a word but simply jerked his thumb indicating that I should get off the bus. I did not realize at the time that this was because he did not give change. I was very mystified by the whole procedure and this was not a good introduction to what eventually became my adopted country. When I eventually got to the Y.M.C.A. this also turned out to be rather bizarre. Each inmate was provided with a bunk in a sort of box and there was a wire screen that could be pulled down and locked. When I told Bill and Ann

Briscoe where I was staying Ann let out a shriek and insisted that I immediately move into Peter Cooper Village with them, which I gratefully did.

I had a very interesting visit to Andre Cournand's laboratory in the old Bellevue Hospital where Bill Briscoe worked. It was here that Cournand and Dickinson Richards did the work on cardiac catheterization for which they received a Nobel Prize. I remember Bill taking me out to lunch and when I tentatively offered to pay he instantly shrugged me off with the statement that he was paid far too much anyway! This statement made something of an impression on me because in London I was paid very little and consequently was as poor as a church mouse. From New York I went to Philadelphia to visit the University of Pennsylvania where Robert Forster and Arthur DuBois were working. I remember that they showed me a mass spectrometer that did not work and had a sheet thrown over the top of it. From there I went on to Baltimore to Johns Hopkins to visit Richard Riley, another giant figure in pulmonary gas exchange. Finding him in the hospital turned out to be a challenge and taught me something about the difference between one's local and international reputation. Nobody had heard of Riley, and I think that even when I found the correct floor of the department he was still unknown. This was amazing to me because he was a household word in our laboratory in London. However I eventually found him.

From Baltimore I flew to Buffalo, New York to visit Hermann Rahn's department. As indicated earlier, I had been greatly influenced by Rahn's work on ventilation-perfusion inequality. I remember that there was a rehearsal in the Department of presentations for the upcoming Fall meeting of the American Physiological Society. I was invited to give my paper which was about our work with radioactive oxygen and it duly impressed them. From Buffalo I went by bus to London, Ontario in Canada where the Fall meeting of the American Physiological Society was being held. This was a wonderful experience and I met many eminent physiologists whom I only knew from their publications. At the end of the meeting I was driven back to Buffalo in a car with a group of other people. There was some sort of meeting in the department of physiology in Buffalo and I stopped off there but the car continued on to somewhere like Baltimore with my suitcase in the trunk. This was something of a disaster and as I moved westwards across the country the suitcase never quite caught up with me until I think I got to San Francisco.

From Buffalo I flew to Rochester, Minnesota where Ward Fowler was at the Mayo Clinic. There was a hiccup at this point because I think the travel agent in England had confused Rochester, New York and Rochester, Minnesota. Ward Fowler kindly offered to have me stay in his home. I remember being amazed at the luxury (it was the first American house I had stayed in) and particularly seeing the children watching television before breakfast. It had never occurred to me that programs would start before midday. From Rochester I flew to San Francisco and was met at the airport by Ted Finley. The people in Rahn's department had contacted him because he had previously worked there. His first greeting was "I hear you dig VAQ" (he was referring to the ventilation-perfusion ratio). He took me to where he was living and said I had a choice of I think 30 bedrooms. It turned out that he lived in a condemned hotel or something. We then went for lunch at The

Cliff, a restaurant overlooking the Pacific Ocean. I was enormously impressed by everybody I met in the U.S.

From San Francisco I flew to Australia on a piston engine Lockheed Superconstellation. The plane was very slow and the drumming of the engines caused ringing in my ears which lasted for several days as I recall. Because the flights took so long we were given accommodation en route. At Honolulu I was put up for a day in a beautiful hotel on Waikiki beach and there was a buffet lunch. I had never seen such marvelous food in my life. We came down again at Christmas Island which was jointly owned by the U.S. and Britain. Its only function was as a landing strip and refueling station. There was a post office where the counter was divided in the middle and you could get U.S. stamps on one side and British stamps on the other. Before we took off the ground crew all lined up and saluted. After a further stop in Fiji I eventually reached Sydney very travel worn. I immediately sent a telegram to me parents asking them to “kill fatted calf.” Of course it was wonderful to see them and be in Adelaide again.

On the return trip I flew from Sydney to Los Angeles and en route was able to spend a few hours with Rosemary who was working in Fiji as a doctor. (She had one of the shortest postal addresses in the world—Dr West, Ba, Fiji). In L.A. I visited Ross Adey at U.C.L.A. He had become a prominent neurophysiologist and I knew him from Adelaide where he lectured us in anatomy. One of the most memorable things I did in Los Angeles was rent a car at the airport and drive down to Marineland. I had never driven a car with automatic transmission before and it took me a few minutes to work out what to do. But I found driving on the big L.A. freeways enormously exhilarating. Then back to New York and on to London. All in all it was a wonderful trip and I think it convinced me that one of these days I would move to the U.S. In fact shortly after my return that I went along to the U.S. Embassy in Grosvenor Square and put my name down for an immigrant visa. In those days this involved a long wait if you were born in Australia.

Chapter 5
Silver Hut Expedition, 1960–1961

I had been elected a member of the Physiological Society, which was considered something of an honor in those days, and on several occasions I had presented our work with radioactive gases at various meetings. During one of these, probably in early 1960, I happened to be sitting next to Maureen Young, a physiologist interested in fetal hypoxia, and she remarked to me "Did you know that Griff Pugh is arranging a medical research expedition to the Himalayas?" I vaguely knew Griff (Fig. 5.1) who had been the physiologist attached to the British expedition that made the first ascent of Mount Everest in 1953, but I had no special interest in high-altitude physiology although I knew something about the topic because of its links to respiratory physiology. This chance remark prompted me to contact Griff, and when we met I explained that I was trained in respiratory physiology, was supported by the Medical Research Council who were financing the scientific part of the expedition, was young, fit, and keen to go. To my surprise Griff accepted me despite the fact that I had never been above an altitude of about 5000 ft in my life. This was the fortuitous beginning of one of the most colorful episodes in my life.

Of course Ed Hillary as leader of the expedition wanted to interview everybody before taking them on. I met him in London and to be honest cannot remember how the interview went. The anecdote I always relate is that he asked me to climb a flight of stairs and when I was successful pronounced me a member of the expedition. I have told this story so often that I am beginning to believe it!

Griffith Pugh was a physiologist who worked in the Medical Research Council Division of Human Physiology in Hampstead in the north of London, and by 1960 he had something of a reputation. He was an accomplished skier and it is said would have represented Britain in the 1936 Olympic Games in a cross-country event if it had not been for an illness. During the War, Griff had trained troops in mountain warfare in Lebanon. Later he was a house physician at Hammersmith Hospital for a period but then joined the M.R.C. Division of Human Physiology.

In 1952 Eric Shipton who was planning an expedition to Cho Oyu just west of Everest asked Griff for advice on oxygen equipment. Griff subsequently took part in the expedition and made extensive studies of the physiological aspects of

J.B. West, *Breathing on the Roof of the World*, Springer Biographies,
DOI 10.1007/978-1-4939-7122-0_5

Fig. 5.1 Lewis Griffith Creswell Evans Pugh (1909–1994), scientific leader of the Silver Hut expedition, with three of his physiologists. *Left-to-right*: Sukhamay "Larry" Lahiri, J.B.W., Pugh, Jim Milledge. This photograph was taken in 1971

climbing to extreme altitude. These are described in my book *High Life*.[1] In fact the knowledge gained on this expedition was critical for the success of the first ascent of Everest by the British expedition in 1953. During both the Cho Oyu and 1953 Everest expeditions Griff, got to know Ed Hillary, and the two subsequently spent further time together during another expedition when Vivian Fuchs and Hillary crossed Antarctica. Apparently at that time they discussed the possibility of a Himalayan expedition which would combine medical research and mountaineering. The result was the Himalayan Scientific and Mountaineering Expedition of 1960–1961 which has subsequently always been known as the Silver Hut Expedition.

The unusual design of the expedition was in three parts.[2,3] In September 1960, a large group walked in from Kathmandu, the capital of Nepal, to the Mingbo Valley just south of the Everest region carrying the parts for the pre-fabricated hut in which the scientists were to spend the winter. The hut was painted silver (Fig. 5.2), hence the popular name for the expedition. This group was led by Norman Hardie, a New Zealand mountaineer, and included Jim Milledge (see Fig. 5.1) from the U.K.,

[1]West, J.B., *High Life: A History of High-Altitude Physiology and Medicine*, New York: Oxford University Press, 1998.

[2]Hillary, E.P., and Doig, D., *High in the Thin Cold Air*, Garden City, NY: Doubleday, 1962. The first part of the book, by Doig, mainly concentrates on the yeti hunt. The second part, by Hillary, briefly describes the early part of the expedition and then deals with the attempt to climb Makalu.

[3]*National Geographic* magazine, October 1962, Vol. 122, pp. 502-551. There are two articles. The first, by Barry Bishop, is a general description of the expedition. The second, by Hillary, describes the construction of the school in Kumjung which took place immediately after the expedition.

Fig. 5.2 The Silver Hut at an altitude of 5800 m (19,000 ft)

Barry Bishop from the National Geographic Society in Washington DC, Walter Romanes, another New Zealander who was skilled in building construction, and some 300 porters. This was a challenging carry because many of the loads were large and awkward (Fig. 5.3), the monsoon rains were still in full swing, and the party had to cross a 4600 m (15,000 ft) pass because bridges were washed out. Nevertheless everything arrived safely and the hut was erected just below the Ama Dablam col at an altitude of 5800 m (19,000 ft) in early November (Fig. 5.4).

While this large party was making their way into the Everest region, another group headed for the Rowaling Valley to look for the Abominable Snowman or yeti. This group included Edmund Hillary and several biologists. This harmless diversion was apparently the brainchild of the Field Enterprises Educational Corporation of Chicago, publishers of *World Book Encyclopedia,* who financed the whole expedition. Needless to say no yeti were found but the search is entertaining described by Desmond Doig in *High in the Thin Cold Air* (see footnote 2).

With the approach of winter, the first group of climbers left and the wintering group of physiologists began their research program. The main site was the Silver Hut at 5800 m (19,000 ft) but measurements were also carried out at the Mingbo base camp, altitude 4650 m (15,300 ft). There was another structure called the Green Hut at about 5300 m (17,400 ft) which was placed there as a fallback position in case 5800 m proved to be too high an altitude to be tolerated during the whole of the winter period.

The third phase of the expedition began in the spring of 1961 when the climbing party returned for an assault on Makalu, the fifth highest mountain in the world at 8481 m (27,825 ft). Over 200 loads were carried across a high level route through

Fig. 5.3 Parts of the Silver Hut being carried in during the fall of 1960

Fig. 5.4 Erection of the Silver Hut. Hillary is in the *center* and *left-to-right* in the background are Mulgrew, Romanes, Gill, and Milledge (*back view*)

the Hongu region to the Barun Valley southwest of Makalu The main objective here was to reach the summit of Makalu without supplementary oxygen although in the event some important scientific studies were also carried out in this third phase.

In preparation for the scientific studies I spent a number of days at the Hampstead laboratory during the summer of 1960 doing exercise tests at sea level so that the results could be compared with those at high altitude. I got to know Jim

Milledge and also saw a lot of Griff Pugh. It was clear that he had some unusual habits. For example he would often arrive to start the day's work at around 10:30–11:00 a.m., stop for lunch around 3 p.m. (much to the annoyance of the kitchen) and then work on to 7 or 8 in the evening. One of my projects was to make control measurements of the diffusing capacity for carbon monoxide in preparation for the high altitude studies. I also spent a day in Dan Cunningham's lab in Oxford for my sea level measurements of the control of ventilation.

I did not walk in with either of the two groups referred to in the description of the expedition above. Instead the plan was for Mike Ward and myself to bring in some additional pieces of scientific equipment in November and December in time for the winter research period. According to my expedition diary which I still have, it was November 19 when I took a taxi from the Kelmscott House flat to Hammersmith Hospital to pick up some crates of equipment that had been delivered there. Then on to Victoria where I said goodbye to Joanna, my girl friend at the time, and teamed up with Mike Ward. Mike was a surgeon who had been on the 1953 Everest expedition. I think we must have taken a bus to Heathrow Airport. The start of the long flight to Delhi with stops in Frankfurt, Beirut, and Bahrain was not propitious because I had a corneal ulcer in an eye and was airsick twice (that travel sickness gene again). However Mike's medications soon fixed the ulcer. We arrived at Delhi Airport at 7 a.m. on Sunday the 20th, piled a large number of wooden packing cases into a couple of taxis, and checked into the very elegant Maidens Hotel. This was my first visit to India and I was most impressed by the boys in white uniforms running about everywhere to do one's beck and call. Mike and I spent the day sightseeing in Delhi which he knew from the celebrations following the 1953 expedition.

Our flight to Kathmandu took off at 7:30 a.m. the following morning and I was suitably impressed by the magnificent views of the Himalayas. We were met by Jim Milledge's wife, Betty who had arranged to work as an anesthesiologist at Shanta Bhawan Hospital in Kathmandu while we were in the Everest area. We stayed in the ostentatious but run down Royal Hotel which had previously been a palace of one of the Ranas, the ruling class. The electric lights kept flickering out and the hygiene was very suspect because most guests developed diarrhea. The next few days were frustrating because of problems finding the air-freighted equipment which was supposed to be in Kathmandu well ahead of us, and getting various permissions from sundry officials. However I met some interesting people including the legendary Colonel Jimmy Roberts who started the whole holiday trekking business with his firm Mountain Travel. I also spent an evening at Shanta Bhawan Hospital with Dr. Bethel Fleming whose son was an expert on Nepalese wildlife, particularly birds.

We eventually set out from Kathmandu on November 30 taking a jeep a few miles to Banepa. From then on it was walking all the way and although this was very tiring at first I soon became very fit. The first day's walk started at 6 a.m. down a hill and on to the Hukse Plain but by the time I got to Dolaghat that evening I was utterly exhausted. I took it more slowly the following day and felt much better. That evening we met Marlin Perkins, director of the Lincoln Park Zoo in Chicago and the host of a well-known television program. He was returning from the yeti hunt which sounded like a bit of a farce. They had set up "Capchur" guns which were supposed to fire hypodermic syringes filled with a tranquilizer into any animal that activated a trip wire but alas no yeti were seen. During the following day much of our walk was along a high ridge with superb views of the Himalayan peaks to the north and we ended up at the Buddhist monastery at Risingo where four rivers meet. During the day we met Ed Hillary, George Lowe, Pat Bartram, Desmond Doig and several village elders from Kumjung. They were on their way out with an alleged yeti scalp which was to be taken around the world by Khunjo Chumbi, the chief village elder of Kumjung. The tour included an audience at Buckingham Palace. Chumbi's comments on the West which are reported in the book by Hillary and Doig (1962) make entertaining reading. Mike Ward vaccinated the nervous Chumbi in preparation for his overseas trip.

The trek continued for the next 12 days from Manga Deorali, Karantichap where Sherpa Nima cooked a chicken which he had carried alive on the top of his backpack all day, Yarsa, the village of Those where there is an enormous chain bridge of great iron links, Chyangma, Sate, Junbesi, Taksindu where there was a splendid monastery, Jubing, an unnamed campsite, Ghat, where we met many Tibetan refugees on their way to Kathmandu, and finally Namche Bazaar. It is interesting that my diary states that I was very disappointed in this major Sherpa village because it was so dry and dusty. By contrast 21 years later the village made a great impression on me. The next day I walked on alone to Thyangboche and was greatly impressed by its magnificent location with the view of Everest peeping over the Nuptse wall. I stopped that night in Changmatang where the expedition had rented a house. Finally on December 17 I reached Mingbo, the site of our base camp (Fig. 5.5) at an altitude of 4650 m (15,300 ft). I remember the final pull up to Mingbo as tough going and I was puffing like a steam engine, but of course it was very exhilarating to reach the end of the long arduous trek.

I had walked in with some of the delicate equipment that we were going to be using for physiological studies including a carbon monoxide analyzer and, most worrying of all, an oximeter which included a small mirror mounted on a long delicate quartz fiber. Both instruments, particularly the carbon monoxide meter, were damaged during the trek in but I had studied them extensively before we started and managed to repair both. In fact by December 20th I had started exercise tests.

The final ascent from Namche Bazaar at 3400 m to Mingbo at 4650 m, that is an increase of 1250 m (over 4000 ft), was done in only 2 days and not surprisingly I had some acute mountain sickness. In particular I slept badly with breathlessness, feelings of claustrophobia on several occasions, and a severe headache. In retrospect it is odd that we did not spend more time acclimatizing at Namche but in those days the relations between ascent rate and mountain sickness were not as clearly understood as they are now.

A week after we arrived at Mingbo we decided to go up to the Silver Hut for Christmas. On the 24th I trekked up to the Green Hut (5300 m) and found the going very tiring with lots of boulder-hopping, and I had severe periodic breathing during

Fig. 5.5 Base camp at Mingbo (4650 m; 15,000 ft). Ama Dablam is behind

the night. Then on the 25th it was up to the Silver Hut for the first time, a very memorable day indeed. My diary notes that we called up the people in the Silver Hut on the radio at 8:30 a.m. to be answered by singing of *O Come All Ye Faithful*. There was an impressive climb up through the icefall and of course a great reception when we arrived. In the afternoon we went skiing on the glacier with some terrific busters and a very tiring walk back to the hut. There was a splendid Christmas dinner with soup, prawns, roast lamb and yak, Christmas pudding and brandy sauce, followed by whiskey on the rocks. I spent the night in the Silver Hut and then returned to Mingbo for another 2 or 3 weeks before going back to the Silver Hut for the long winter period. My birthday on the 27th was occupied by making measurements of the diffusing capacity for carbon monoxide on various people in the base camp, and my diary notes that I turned 32 which I regarded as almost middle-aged.

December 30 was a red-letter day. I was woken in the morning by Mike Ward shouting "helicopter" and sure enough one landed just outside our mess hut. The pilot was a very pleasant Swede who carried no oxygen in spite of the altitude. One of the passengers was from the Red Cross investigating the plight of the Tibetan refugees. On January 5 the helicopter reappeared this time with Ed Hillary and Desmond Doig. Ed entertained us in the evening with the story of his round the world trip with Chumbi and the yeti scalp.

Once I had acclimatized to the altitude at Mingbo I enjoyed the period there. Each morning I would go for a walk, and there were spectacular views of Ama Dablam (see Fig. 5.5) and, about 4 km to the north, Everest peeping over the Nuptse wall. The nights were always warm enough in my cozy sleeping bag but the mess hut was cold because its walls were simply stones piled on top of each other with a plastic sheet over the top. As a result the evening meal was a bit miserable. Griff Pugh had his own private tent with a charcoal brazier but one of the hazards of this was carbon monoxide poisoning. I was making measurements of carbon monoxide in expired breath in order to measure the diffusing capacity of the lung. At one stage when we were in the Silver Hut I found to my astonishment that the subjects were exhaling more carbon monoxide than I was giving them to inhale! The explanation turned out to be that the Sherpas had been using primus stoves in the Hut for cooking and everybody who was living in it had absorbed carbon monoxide from the air. Of course we removed all open stoves from the hut after that.

By January 9 I had finished the first series of measurements of carbon monoxide diffusing capacity at Mingbo and on the 12th I left for the Green Hut and then the Silver Hut. To most people who have not spent time at very high altitudes there probably seems little difference between the altitude of our base camp at Mingbo (4650 m; 15,300 ft) and that of the Silver Hut (5800 m; 19,000 ft). In fact however there is a world of difference between these two altitudes in terms of their effects on the body. Essentially all of us were able to acclimatize satisfactorily to the Mingbo altitude after a couple of weeks or so, but living in the Silver Hut was a much greater challenge. At that altitude you are constantly aware of shortness of breath and even a small effort such as sitting up in the bunk at night leaves you panting for breath for a couple of minutes. There was a marked loss of initiative so that it was always difficult to start new projects. Memory was poor and people made a lot of arithmetical and other errors. We found that we could carry out experiments if the protocol was clearly defined and especially if there was a checklist, but problem solving when there was an unexpected situation was slow and difficult. Appetites were poor and there was a striking weight loss of 0.5–1.5 kg (1–3 lbs) per week in all expedition members living in the hut. Some people regained weight if they descended to the base camp and one of the remarkable findings was that on going down the 1150 m (3800 ft) of altitude resulted in an enormous improvement in mood and appetite. As far as we knew, no one had ever spent such a long period at such a high altitude as the Silver Hut before.

The Hut provided remarkably comfortable living quarters during the winter in spite of the intense cold outside. It had been prefabricated in England and consisted of a number of curved panels of marine plywood with plastic foam insulation. This gave it the shape of a cylinder with a flat bottom about 6.7 m long and 3 m wide and it was painted bright silver hence its name. The hut was divided into three parts. At

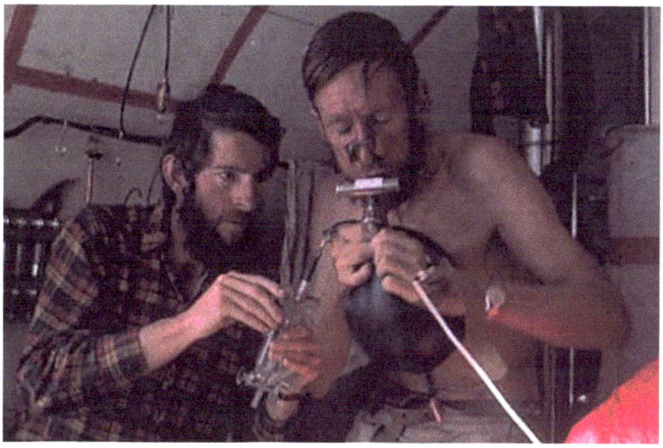

Fig. 5.6 Jim Milledge measuring the cardiac output of myself during exercise on the cycle ergometer in the Silver Hut. The uptake of acetylene by the pulmonary blood flow gave the cardiac output

one end were eight bunks arranged in upper and lower levels. Near them was the door with a porch so that people did not track snow into the living area. In the center of the hut was a kerosene stove which worked so well that the usual complaint was that the hut was uncomfortably hot. The kerosene drained into the stove by gravity from tanks outside the hut and a flue led combustion gases to the outside. The far end of the hut contained the laboratory with a stationary bicycle for exercise measurements and extensive physiological equipment (Fig. 5.6). There was a memorable day in June 1960 when I drove down to Maidenhead to help with a trial of assembling the Silver Hut which had been constructed by a building firm there. The assembly went very rapidly under Ed Hillary's direction and afterwards we were all invited to tea at John Hunt's house in Henley-on-Thames. (Hunt was the leader of the British expedition that first reached the summit of Everest in 1953.) I had my friend Joanna with me and she was suitably impressed.

There were seven physiologists working in the Silver Hut during part or all of the winter including Griff Pugh (scientific leader), Mike Gill (New Zealand medical student), Larry Lahiri (Indian physiologist), Jim Milledge, Tom Nevison (physiologist with the U.S. Air Force), Mike Ward (London surgeon) and myself. Five of these are shown in Fig. 5.7. In addition Barry Bishop, a geographer from the National Geographic Society, and Wally Romanes, a builder, were present.

We did not know what kind of weather to expect during the winter. Of course we knew it would be cold but we had little idea about how much snow there would be. In the event there were a few storms but little precipitation. Most of this occurs during the monsoon season in the summer. In fact towards the end of the spring when the Silver Hut was finally taken down, it was perched on ice stilts because the general level of the snow and ice had been so ablated by the sun.

We had a dog in the Silver Hut for a period. This was a Tibetan terrier called Rakpur and it had followed someone up to Mingbo from one of the Sherpa villages. However the dog did not tolerate the altitude well and was very listless in the hut. On one memorable occasion it peed on Tom Nevison who summarily threw it out the door.

Fig. 5.7 Five of the physiologists on the glacier outside the Silver Hut. *Left-to right*: Jim Milledge, J.B.W., Griff Pugh, Mike Ward, and Mike Gill

Without going into too much detail, the main aim of the physiological program was to elucidate the changes that occur in lowlanders (people who normally live near sea-level) when they spend an extended period at a very high altitude. Many of the studies concerned exercise because it was well known that the maximal exercise capacity was greatly reduced at these altitudes and we wanted to elucidate the reasons for the limitation. Measurements were made of the oxygen levels of the blood during heavy exercise and we showed that these levels dramatically declined as the exercise level increased because the oxygen was not able to diffuse rapidly enough from the lung into the blood. One of my main projects was to measure the diffusing capacity of the lung because it was clear that this was one of the chief limiting factors for working at high altitude. The results showed that the diffusing capacity was almost unchanged, that is the lung did not adapt its structure to a long period of high altitude. Other measurements included the factors controlling the level of ventilation at high altitude, changes in the composition of the blood, the electrocardiogram, and brain function. Cognitive function was studied by using a card-sorting test and it was found that cards could be sorted accurately at this very high altitude but this took longer. There were also renal, metabolic and endocrine studies. The physiological program is summarized in Chap. 11 of my book *High Life*. The Silver Hut expedition was by the far the most ambitious and successful high-altitude expedition of its time as judged by the overall scientific return.

Of course life in the Silver Hut was not all work. We generally spent the morning and early afternoon carrying out experiments and then would go skiing on the glacier, weather permitting (Fig. 5.8). The expedition had been given the latest

Fig. 5.8 Jim Milledge took this beautiful photograph of me skiing on the Mingbo glacier. The peak behind me is Numbur which is sacred to the Sherpas

metal/plastic Head skis that were very much an innovation at that time. Unfortunately the bindings had not been fitted and so we did this ourselves using old "bear trap" bindings which had little or no emergency release capability. On one occasion Jim Milledge went barreling down a hill, hit soft snow at the bottom, and was thrown head over heels. We expected a couple of broken ankles but instead the bindings were torn out of the skis! Skiing down the glacier was fun although of course the snow was completely ungroomed and occasionally was crusty on top which was dangerous for amateurs like ourselves. However the most memorable part was the tramp up from the bottom carrying your skis. The evenings were spent reading and talking and we also had occasional radio connections with an American firm in Kathmandu who were putting in a communications system. I often ran the radio which used the amateur bands and our call sign was 9N3PM. The prefix 9 N3 for eastern Nepal was recognized by radio hams who heard it as extremely rare, possibly unique, and we often had requests for contacts with them.

At one stage during the winter, Barry Bishop, Mike Gill, Wally Romanes and Mike Ward decided to try to make the first ascent of Ama Dablam a spectacular peak 6856 m (25,500 ft) high that dominated the skyline to the northwest (see Fig. 5.5). We were able to watch their progress through a telescope and it was a very tough climb taking 3 weeks or so but they were ultimately successful. At one point one of the Sherpas who was helping with the climb broke his leg when an unstable boulder rolled on to it. Carrying him off the mountain was a major challenge. The Nepalese authorities claimed that the expedition did not have permission for this ascent and they were very angry and threatened to cancel our attempt on Makalu. As a result Ed Hillary had to take a helicopter back to Kathmandu and bow and scrape to various officials and pay a substantial fine before we could continue.

One of the unusual accomplishments of the expedition was to build what was one of the highest airstrips in the world. This was a little above our base camp at an altitude of about 4700 m (15,400 ft). The reason for this was that the International Red Cross was anxious to assist the large number of Tibetan refugees who were streaming over the border and who were short of food. Ed Hillary selected the site and it was cleared by Sherpas who were grateful for the extra income. A logistical problem was the presence of two huge boulders on the runway weighing perhaps 20 tons each and protruding 6 ft above the ground. The Sherpas solved this by digging enormous craters beside each of the boulders and then tipping them into these with powerful levers. I remember the first flight in when the Pilatus Porter plane pancaked and broke its tail-wheel and a makeshift repair had to be carried out. I also vividly remember the plane taking off. It roared down the runway and then suddenly disappeared as it dropped into the valley below eventually acquiring enough airspeed to climb out. As the plucky Swiss pilot, Captain Schrieber, said the airstrip was not for the nervous.

My diary had a long lapse in February and petered out completely on March 5, an indication of the loss of drive that occurred at this very high altitude. However towards the end of April a group of us left the Silver Hut and ascended the very steep 500 ft climb to reach the Ama Dablam col. We were now on our way via a high-altitude route to Makalu This involved traversing the Hongu region and crossing two more high cols before we dropped down into the Barun Valley where the Makalu Camp 1 was located. By this time I had spent almost 4 months at an altitude of 19,000 ft and was wiry and tough. Ed Hillary in his book *High in the Thin Cold Air* describes how when he was crossing this route which goes up to 6100 m (about 20,000 ft) in places, Jim Milledge and I who were with him were moving faster than he found comfortable!

In the initial planning of the expedition, not much thought had been given to possible physiological experiments on Makalu. However the winter scientific program had been so productive, and several of the physiologists, particularly Mike Gill, Jim Milledge, Mike Ward and myself seemed so fit, that Griff Pugh agreed that we should try to do some measurements at even higher altitudes on Makalu. Griff himself stayed behind at the Silver Hut or went down in Mingbo.

Of course it was enormously exhilarating for me to find myself on the fifth highest mountain in the world. George Mallory had described it thus "among all the mountains I have seen, and, if we may judge by photographs, all that ever have been seen, Makalu is incomparable for its spectacular and rugged grandeur." This was quite an introduction to mountaineering for someone like myself who had never done any before. Reaching the base camp at Mingbo was simply a trek although the last stages left me pretty breathless. The climb from Mingbo to the Silver Hut through an icefall involved some technical difficulties and of course the altitude was challenging but it was a well-traveled route. Makalu on the other hand was very special.

Camp 2 on Makalu at 5940 m (19,500 ft) was already higher than the Silver Hut, and Camp 5 on the Makalu Col was at the extreme altitude of 7440 m (24,400 ft). The climbers doggedly made their way up the mountain in spite of a great deal of very cold and windy weather. When Jim Milledge and I reached Camp 3, altitude 6400 m (21,000 ft), we set up the stationery bicycle at and made some measurements of maximal work and also recorded some electrocardiograms. While we were there one of the Sherpas, Aila, developed some kind of lung condition which I thought was pneumonia, but subsequently Jim Milledge told me that he recognized it as high-altitude pulmonary edema which had only been clearly described in the English literature a few months before. Jim and I shepherded Aila down to Camp 2 while he coughed up pink frothy fluid!

At Camp 2 an event occurred that had a very serious effect on the final stages of the expedition. Ed Hillary was in his tent and he had been complaining of pain on the left side of his forehead for 2 or 3 days. He had been as high as Camp 4 at 7010 m (23,000 ft) but the severe headache and a general feeling of grogginess had forced him down to Camp 2. Late in the afternoon Jim Milledge and I were in our tent when we heard strange noises from Ed's tent, and it turned out he had lost his ability to speak sensibly although he was perfectly rational in his thinking. Clearly he had developed some sort of cerebral vascular abnormality. Mike Ward and Jim took control and insisted that Ed go down to a lower altitude as soon as possible and return to Kathmandu via a low level route. Jim Milledge agreed to be the physician to accompany Ed, and of course this was a disappointment to Jim because it meant that he was not able to take any more part in the science or climbing at extreme altitudes.

Some days later I reached Camp 5 on the Makalu Col where I spent several days. It was a spectacular location on a saddle right on the border between Tibet and Nepal. Everest, Lhotse and the South Col dominated the skyline only a few miles to the west (Fig. 5.9). The photograph shows Mike Ward and myself assembling the stationary bicycle with which we made measurements of our maximal oxygen consumption. To do this we needed to know the composition of the air that we exhaled. The air was collected in a large gas-tight bag, and samples were then taken in pre-evacuated glass ampoules which were flame-sealed and analyzed back in London when the expedition returned. My maximal oxygen consumption was 1.46 liters per minute, that is about a third of my sea level value. Mike Ward's maximal oxygen consumption was slightly less probably because he had some high-altitude pulmonary edema. For some time after the expedition I used to state on my curriculum vitae that I held the world record for measured maximal oxygen consumption at 7440 m! When pressed I would concede that it was only measured on two people and that the other one had pulmonary edema. I also recorded 12 lead electrocardiograms at Camp 5, and in addition we collected alveolar gas samples, again using the sealed glass ampoules that were later analyzed in London. In fact Mike Gill collected alveolar gas samples as high as 7830 m (25,700 ft), a remarkable accomplishment. Incidentally I remember a nasty accident in our tent at Camp 5 when we were attempting to seal the glass tubes. This was done using an oxypropane torch and the propane must have leaked. There was a loud whoof and several eyebrows and a sleeping bag were burned. Of course at that altitude the margin of safety for almost everything is very low.

Fig. 5.9 Mike Ward and myself assembling the cycle ergometer at Camp 5 on the Makalu Col (7440 m; 24,400 ft). The resulting measurements of maximal work capacity are still the highest reported field measurements to date. In the background is the summit of Mt. Everest (*right*), and to the *left* of that is the South Col and Lhotse seen from the east

Makalu had been previously climbed by a French expedition in 1955 but they had used large amounts of supplementary oxygen, and the plan of our expedition was to try and make an ascent without oxygen. One of the hopes was that the exceptionally long period of acclimatization of the wintering party would result in high levels of performance on the mountain. However in the event the expedition was not successful in reaching the summit. Pete Mulgrew and Tom Nevison reached an altitude of about 8352 m (27,400 ft) according to their altimeters when Pete suddenly collapsed gasping for breath. The most probable explanation was that a clot of blood had been swept into his lungs resulting in a pulmonary embolism. All thought of climbing the mountain immediately disappeared and the next few days were devoted to a desperate attempt to get Mulgrew off the mountain alive.

With great difficulty he was eventually brought down to Camp 5. At this time I was in Camp 3 and was asked to help. We had some rescue oxygen and I used this to assist me in climbing back to Camp 5. The effect of the oxygen was nothing short of miraculous. I steamed up the mountain well ahead of the Sherpas who were accompanying me and if anybody claims that supplementary oxygen is not of great value at high altitude, don't believe them. It made an enormous difference to my ability to climb. When I got to Camp 5 it was clear that the situation was desperate. Mike Ward was there but hopelessly confused. In retrospect it is likely that he had both high altitude pulmonary and cerebral edema. Pete Mulgrew was unmanageable. Although we tried to assist him down the mountain he struggled and was incoherent. I finally gave him an injection of morphine which quieted him, we strapped him to an improvised stretcher made of backpack frames and poles, and bundled him down the mountain. The route was very steep and difficult and he was unconscious most of the way; indeed I checked his pulse from time to time to see whether he was alive.

Eventually Pete was carried down to Camp 1 strapped on to the back of a Sherpa (Fig. 5.10) and he was then helicoptered back to Kathmandu. Unfortunately he had to have both of his lower legs amputated because of severe frostbite and he developed an addiction to pain-relieving drugs. However he was eventually fitted with prostheses and became a successful businessman and accomplished yachtsman in Auckland.[4] Tragically he lost his life on the Air New Zealand flight which crashed into Mount Erebus during a sightseeing visit to Antarctica. Pete and Ed Hillary alternated as narrators for these flights. Parenthetically it is remarkable how

Fig. 5.10 Pete Mulgrew being carried back by one of the Sherpas

[4]Peter Mulgrew's own graphic account can be read in Mulgrew, P.D., No Place for Men, Wellington, NZ: Reed, 1964.

many high-altitude mountaineers meet violent deaths though not necessarily on mountains. In addition to Pete Mulgrew on this expedition, Barry Bishop lost his life when his car ran off the road at night, Lee Ortenberger was killed in a fire in Oakland, California a few years ago, and John Harrison died in an avalanche.

I had developed severe hemorrhoids during the last stages of the expedition (a common problem at high altitude) and was dreading the long walk out because of the pain. However we had two other people in addition to Pete Mulgrew who needed evacuation by helicopter including Mike Ward who was recovering from high-altitude pulmonary edema, and Ang Temba who had a broken ankle. It was a great relief to be lifted from Camp 1 at 5500 m (18,000 ft) and taken down with Ang Temba to the remote village of Num at 4000 ft on the banks of the Arun River. I remember that the helicopter had great difficulty in taking off at that high altitude and was only able to do this when there was a puff of wind. The following day I was in Kathmandu.

Following the expedition I made a brief visit to Australia to see my parents. There was a problem in that my passport was stored in a snow cave near the Silver Hut. There had been no point in carrying it over to Makalu and of course I expected to return by the same route that I had walked in. However the British Embassy in Kathmandu was very helpful and issued me a British passport which lasted long enough to get me to Australia and back to England. The passport photograph showed that I had lost a great deal of weight and had also grown a scraggy beard. In fact my parents hardly recognized me when I reached Australia.

Chapter 6
University of Buffalo and the Postgraduate Medical School, 1961–1967

When I returned from the Silver Hut Expedition my elegant flat in Kelmscott House was waiting for me. For several weeks after I got back I thought a lot about the expedition and what a remarkable experience it had been. I can remember having lunch by myself on the deck of the pub "The Dove" only a stone's throw from the flat and musing on the expedition over a glass of beer, and being amazed that such an adventure had really happened to me. In fact this period was something of an intermission. Of course I was writing up the results of the scientific studies for publication. However prior to leaving for the expedition in late 1960, I had written to Hermann Rahn (Fig. 6.1) inquiring whether it would be possible for me to spend a year in his laboratory in Buffalo, and he invited me to do this.[1] Consequently I just had a few weeks in London before leaving again to spend another year away. As I indicated earlier there had been some problems in arranging for someone to occupy my flat while I was on the expedition, and I now decided that it would be impossible to keep it for yet another year and I somewhat reluctantly let it go. I later regretted this decision.

I flew to New York and then to Buffalo on the cheapest overnight ticket I could find and remember arriving feeling very sorry for myself. However I was met by both Hermann Rahn and Leon Farhi, which was good of them. Leon was originally from Egypt and like myself had a strong interest in pulmonary gas exchange, particularly ventilation-perfusion relationships. He showed me around the campus and someone helped me find a small apartment close by. There was a visa problem because I had been reluctant to apply for a J-1 visa, which would have been the obvious thing to do. However this would have meant me losing my place in the

[1]Rahn's original research on pulmonary gas exchange and mechanics was carried out with Wallace Fenn and Arthur Otis in Fenn's laboratory in the University of Rochester starting in the 1940s. There is an extensive description of the this in Chap. 2 of Volume I of West, J.B. (ed.), *Pulmonary Gas Exchange; I. Ventilation, Bloodflow and Diffusion; II. Organism and Environment*, New York: Academic Press, 1980. Rahn moved from Rochester to be Chairman of the Department of Physiology at the University of Buffalo Medical School in 1956.

J.B. West, *Breathing on the Roof of the World*, Springer Biographies,
DOI 10.1007/978-1-4939-7122-0_6

Fig. 6.1 Hermann Rahn (1912–1990) whose work on pulmonary gas exchange had a large influence on me. I spent a year in his department at the University of Buffalo

queue for an immigrant visa. Instead I arranged to come on a student visa, which for some strange reason meant that my place in the queue was not affected. In the event this was a good decision because while I was in Buffalo my number came up with the immigration people and I was able to get a Green Card. However a complication was that the University of Buffalo insisted that I had to pay a substantial tuition fee as a student and I could certainly not afford this. Fortunately Hermann Rahn came to the rescue.

I have mixed feelings about my year in Buffalo. The Department of Physiology had some excellent physiologists in it quite apart from Hermann Rahn, including Leon Farhi, Charles Paganelli, and Don Rennie. Also this was my first introduction to a basic physiology group. Most of my previous experience, for example at the P. R.U., had been in a more clinical or human physiology setting. Examples of my ignorance were that I did not know who Paul Bert was (he was the father of low- and high-pressure physiology), or Hans Ussing (who developed the physiology of transport across membranes). As explained earlier my exposure to physiology in Adelaide had been very superficial. I was also chagrined to attend a seminar by a well-known Harvard physiologist who clearly thought that the degree of M.D. rather than a Ph.D. led to a lack of rigor.

Unfortunately the initial experimental project that was assigned to me involving an oxygen electrode was disappointing. In fact the equipment that I had been working with in London was far more sophisticated. The project that I ended up with was also very prosaic compared with the exciting research in London and no peer-reviewed publication resulted from my year in Buffalo. Unfortunately I did not get along well with Leon Farhi for some reason (I was not the only person to have this experience). Initially I was lonely and found it difficult to meet people. This improved when I joined a madrigal choir that one of the technicians in the department introduced me to. Later on I got to know an interesting assistant professor in the music department, Frank D'Accone, and I rented his apartment when he went to Florence to do research in the summer. He was working on Italian renaissance musicians, and he later became a world authority on Italian renaissance music. He subsequently moved to UCLA where he headed a strong program, but unfortunately I lost touch with him and have not been able to re-establish contact. Through Frank I met several interesting people late in my stay.

Buffalo itself was not a stimulating city for a newcomer. I remember that at the time it was losing population and one of the exhortations on the radio was to "Boost Buffalo." I used to listen a lot to the Canadian Broadcasting Corporation, which was closely modeled on the B.B.C. An annoying problem arose in the spring and summer when many people from Buffalo went to Canada during the weekend. However my student visa would not allow me to leave the U.S. and re-enter. Of course hundreds of people were streaming across the bridge from Canada to the U.S. on a late Sunday afternoon and all the immigration officer would do was put his head through the window of the car and ask you where you were born. Unfortunately I never learned to say Boston, Buffalo or whatever with a sufficiently convincing accent!

While I was in Buffalo, Ed Hillary was spending the year after the Silver Hut expedition in Chicago lecturing for the World Book Encyclopedia. He had agreed to do this in return for the World Book's complete financing of the expedition, and he invited me to visit him for a weekend. This was truly memorable. I was entertained by John Dienhart, a big P.R. man for World Book Encyclopedia. He had a mansion outside Chicago by an artificial lake. In the evening he took us to the Playboy Club which was rather new and very much an eye-opener for someone like myself who had been living in Spartan circumstances in Buffalo.

A few months after I had arrived in Buffalo I received an invitation from the Worcester Foundation in Worcester, MA to visit and give them a lecture. I thought this was a bit odd because I did not know the group but they were offering a handsome honorarium and it also gave me an opportunity to drive to Boston and Dartmouth to see other physiologist friends. The invitation came from Gregory Pincus who had been largely instrumental in developing the birth control pill. When I arrived at the Foundation and walked into his office his face fell and it turned out that I was the wrong Dr. West! He thought I was a pharmacologist, and indeed there was another Dr. West who was a pharmacologist in the University of London. However he put a brave face on it and my lecture was about the Silver Hut Expedition which was interesting for everybody including the local women's sewing circle.

A far more stimulating invitation arrived later. This was from Dr. Walter Bauer who was Chairman of the Department of Medicine at the Massachusetts General Hospital. They were looking for a new chief for the Pulmonary Division and invited me to Boston to give a seminar. Meeting Dr. Bauer was bizarre because he had severe lung disease and at one stage in our negotiations was breathing through a tracheostomy tube and had to close this to speak. In any event I gave a seminar on our work with radioactive oxygen and apparently he was impressed. At that time he was looking for faculty who had a very strong research background. When I told him that I wasn't at all sure that I wanted the clinical responsibilities of being chief of the pulmonary division he said that was no problem because another person in the group could handle that side. We left it that I should write an application to the N.I.H. to site a cyclotron at the M.G.H. to do research with radioactive oxygen which I duly did. Since the budget included the purchase and housing of the cyclotron, it was very substantial.

When I returned to London at the end of my period in Buffalo I was amazed to learn that the application had been funded, and of course Dr. Bauer and others at the M.G.H. were anxious for me to take up the appointment. However I now had second thoughts about this. First I was not sure that the arrangements that he suggested for looking after the clinical responsibilities would work. I was worried that I would be saddled with a lot of clinical and administrative issues that I would not feel comfortable with, and I knew that my strengths were in research not clinical medicine. Additionally I was worried about putting all my eggs in the basket of cyclotron-produced isotopes. After all if the N.I.H. were going to provide a cyclotron for me, they were certainly going to expect a long-term commitment to this area. In retrospect I think this was an unnecessary concern, and in fact the use of short-lived isotopes combined with positron-emission tomography (PET) became a very productive field. One of the physicists at the M.G.H., Gordon Brownell, was at the forefront of imaging, and I think that if I had gone back, I would likely have had a productive career although in a very different area.

When John McMichael who was chairman of medicine at Hammersmith heard about the offer he was very anxious not to lose me. He said that he would arrange for me to have my own Medical Research Council respiratory research group with sufficient funding, space and technical help. This was an extremely difficult decision for me. However I eventually came to the conclusion that I would do better by staying at Hammersmith, concentrating on research, and working out the physiology of the effects of gravity on the lung. I will never know whether that was the correct decision. Of course when various people, particularly in the U.S., heard that I had turned down a position as assistant professor at Harvard Medical School, and chief of the pulmonary division at the M.G.H., they were incredulous and probably rightly so. I still sometimes ponder this decision and one comforting thought is that if I had gone to Harvard at that time I would never have met Penelope who became my wife. Also life in La Jolla is very hard to beat. In the event, the M.G.H. went ahead and installed the cyclotron, which I think was only the third in the world specifically designed for medical research, the other two being at Hammersmith and Washington University in St. Louis. I still occasionally meet people who were at the M.G.H. in 1962 and who tease me about this episode.

In the early summer of 1962 when I was living in Frank's apartment in Buffalo, my parents flew to the United States so we could have some time together. I drove my car to New York City where I met them and we saw some of the sights there. We then drove down through Philadelphia and Baltimore to Washington, D.C. and Mount Vernon. It was a great sightseeing trip. The car was an old but very comfortable Chevrolet convertible and on one memorable day we were driving along with the top down when my mother warned us that it looked like rain. Of course my father and I dismissed this airily but the heavens suddenly opened and we were soaked. Later we drove back to Buffalo and my parents stayed with me in my very nice apartment.

Towards the end of my stay in Buffalo a number of the people in the department made arrangements to attend the meeting of International Union of Physiological Sciences in Leiden, Holland. This was my first I.U.P.S. Congress and I have attended many since. They are held every 3 or 4 years. Hermann Rahn explained that the department would pay all my expenses and I was amazed at the generosity. The difference between the U.S. and the U.K. on money matters was dramatic. Incidentally while I was in the department, the University of Buffalo as it then was became part of the State University of New York system, and I had a brief appointment as an assistant professor. I was told that I was welcome to stay in this tenure track position if I wished.

When I returned to Hammersmith we embarked on an extensive research program to elucidate how gravity affected the distribution of blood flow, ventilation and gas exchange in the lung. This period from the fall of 1962 until the middle of 1967 was very productive for me and consolidated my reputation as an experimental physiologist. Much of the work was done using an isolated perfused lung preparation that allowed us to alter the pulmonary vascular and alveolar pressures at will, and determine how the changes affected the distribution of blood flow and gas exchange (Fig. 6.2). One of the best known products of this period was the three

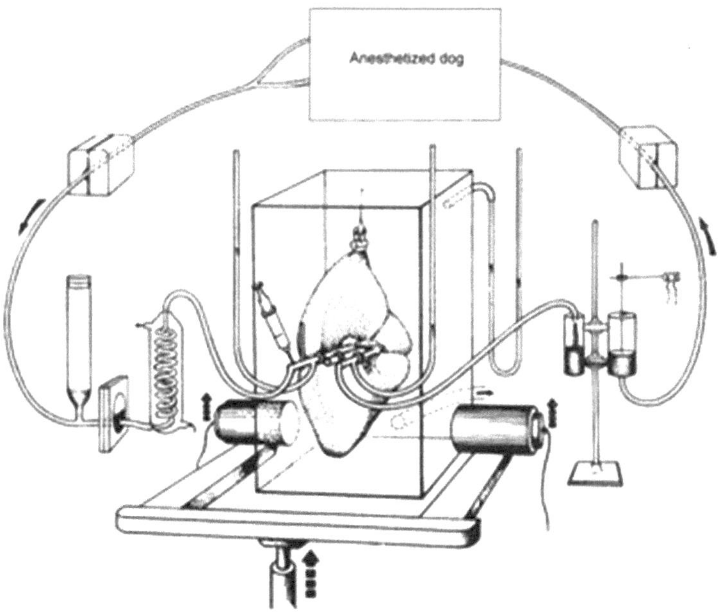

Fig. 6.2 Isolated perfused lung preparation with which we analyzed the mechanisms responsible for the distribution of blood flow in the lung caused by gravity. Radioactive xenon dissolved in saline was injected into the pulmonary artery line and the distribution of blood flow was determined by moving the counters over the lung

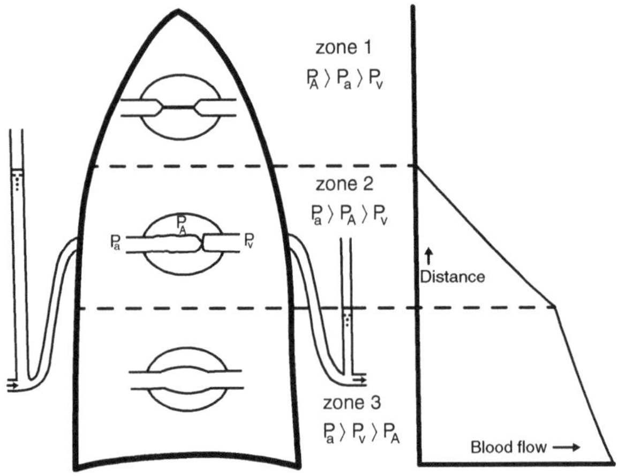

Fig. 6.3 Three zone model of the lung that we developed using the setup shown in Fig. 6.2

zone model of the lung that shows how the relations between the pulmonary artery pressure, pulmonary venous pressure, and alveolar pressure determine the distribution of blood flow in the lung (Fig. 6.3). This diagram is now in many textbooks. We also rapidly froze the tissue on the outside of the lung with very cold liquid Freon to relate the morphology of the capillaries to the physiology. This began a central interest of mine, which is relating structure and function, and this has continued, especially with electron microscopy.

Many excellent postdoctoral fellows and others collaborated with us during this period including Derek Bergel, Kemp Fowler, Jon Glazier, Michael Hughes, Norman Jones, Christine Matthews, John Maloney, Arnold Naimark, Michael Pain, Hans Simon, and Piero Zardini. One memorable experiment was done using the centrifuge at the Institute of Aviation Medicine in Farnborough. Anesthetized animals were frozen by surrounding them with solid carbon dioxide during acceleration over several hours to determine the distribution of alveolar size. Another stimulating study was a modeling analysis of the distortion of the lung under its own weight that was carried out with Frank Matthews from the Department of Aeronautical Structures at Imperial College. This was done using a sophisticated finite element analysis that had been developed by John Argyris to handle large distortions.

As explained earlier, I had to relinquish the flat by the river when I spent the year in Buffalo. Shortly after my return I moved into an elegant flat in a mews near Gloucester Road. This was in a very nice area of London and the flat was very comfortable but it lacked the color of living by the river.

At about this time I was briefly married and there was a child, Julia. However the marriage was very unhappy and I do not plan to say any more about it. Julia later lived in Sydney with a fine family including her husband, Mark, son Miles and daughter Isabelle, and we kept closely in touch. Penelope and I visited them in 2001 and 2015, and Julia and Mark have stayed with us in La Jolla. However Julia tragically died in 2016 from a very aggressive cancer.

Towards the end of this 5 year period when I was in my late thirties I began to think more about my future career. I wondered about the opportunities in London, particularly those at the Postgraduate Medical School, which had recently become the Royal Postgraduate Medical School (R.P.M.S.). Our research program had been very productive and well supported by the Medical Research Council but it was not clear whether there would be a career for me at the R.P.M.S. with my interests in physiology. For example, since the School was exclusively devoted to postgraduate medical studies, there was no opportunity to teach physiology to undergraduate medical students. I therefore made an appointment to visit the M.R.C. head office to ask them specifically about the chances of long-term support. The answer was not encouraging. They pointed out that the M.R.C. had strongly financed respiratory physiology for many years in the Pneumoconiosis Research Unit, and I was told that their interests were moving more towards molecular biology which certainly made sense. However this response led me to wonder whether I should now look overseas. An additional factor was that there was continual grumbling at the R.P.M.S. about academic salaries.

During the middle sixties the U.S. space program was rapidly accelerating and this of course was the decade during which President Kennedy had pledged to put a man on the moon and bring him back safely to earth. Since a major part of my research had been on the effects of gravity on the lung, I thought it would be fascinating to see what happened to pulmonary function in weightlessness. I did not have any contacts in the U.S. space program and it may have been one of my friends at Farnborough who suggested that I write to John Billingham. He had previously been at Farnborough but had emigrated to the U.S. and was now working with NASA at the Ames Research Center in the San Francisco Bay area. Since my earlier visits to the U.S., I had always thought that California would be the place to end up in if possible. The climate and outdoor life reminded me very much of Australia and I liked the friendly "can-do" attitude of the people. I remember watching a television program during a drab, wintry day in London and seeing an English astronomer driving up to the Palomar Observatory in southern California in an open convertible car and I thought that this looked very attractive. So I wrote to John Billingham and he replied that I would be welcome and that he could arrange a very attractive National Science Foundation fellowship for me for a year. Incidentally I did think of writing to Julius Comroe at the Cardiovascular Research Unit in San Francisco and probably he would have found a place for me too. Academically this would have been a safer bet, but I thought that getting a foot in the door in the space program would be very exotic. Oddly enough John Maloney, another Australian, who was working with me at Hammersmith also independently decided to spend some time at the Ames Research Center.

During this 5 year period in London I continued to play tennis at the Holland Park Club and represented the Club in various matches. On one occasion we hosted the Hurlingham Club for mixed doubles and one of my opponents was Penelope Banks who immediately attracted my attention. Also at about this time I became a member of the splendid Hurlingham Club partly because several members of the Holland Park Club also belonged it. The Club has to be the most attractive in

Fig. 6.4 Main building of the Hurlingham Club in London. Interestingly, the original structure, subsequently expanded, was built in 1760 by William Cadogan, a famous London physician and friend of Garrick and Sheridan

London with beautiful gardens, a very fine clubhouse with some eighteenth century rooms and excellent sports facilities (Fig. 6.4). At that time it was not difficult to join with sponsorship from my friends at Holland Park, and I clearly remember the day when my application was accepted and I walked around the grounds feeling very pleased with myself. I might add that it soon became difficult to join the Club because it was so popular, and today it is almost impossible. Those were the days when the club had informal dances outside during the summer and I attended several of these. Of course I played tennis at Hurlingham and also squash. In fact I had been playing squash for a number of years and for a period belonged to the Grampians Club on the south side of Shepherds Bush Green where one of the Khans from Pakistan was the pro. The Khan family dominated world squash for many years.

At Hurlingham I had good opportunities to see Penelope from time to time and I became very fond of her. It was particularly nice to drive down with her to her home, Beech Coppice, some fifteen miles south of London and meet her parents. Penelope herself shared a rather modest flat near Victoria where the skylight leaked when it rained and I always claimed that I could provide something better. Having said this, when I was getting to know Penelope I drove a terrible old car in which the windscreen wipers did not work, and there was some sort of fungus growing on the floor. My salary at the Postgraduate Medical School was always meager but I think that driving a battered old car was part of the academic image. Incidentally I had

had two or three secondhand cars in London, usually convertibles that I drove with the top down all winter except when it was pouring with rain. One was a very sporty cream MG TD and I am sorry that I disposed of this.

Penelope and I agreed to marry but that there was a problem in that I had committed myself to going to California for a year. Her parents were obviously not very pleased to hear that. However Penelope's mother was an American from the New Jersey/New York area and so Penelope had many American relatives. She had actually worked in New York City for a time as a secretary in the United Nations Organization. Therefore it was not as difficult for her to spend some time in the U.S. as it would have been for many English girls. I still had a "green card" although I had had some problems in keeping this while living in the U.K. and only visiting the U.S. occasionally. There were no visa difficulties in Penelope coming with me for a year.

We were married on October 28, 1967 and left for California shortly after that. It was a splendid wedding at St. Andrew's Church in Kingwood, Surrey where Penelope had grown up (Fig. 6.5). Michael Hughes was my best man, and there

Fig. 6.5 Photograph taken at the wedding of Penelope and myself

was an elaborate reception in a marquee on the Beech Coppice lawn just adjacent to the church. Unfortunately my parents could not attend but I had many friends from my Hammersmith days. By this time I had bought a small house in Denbigh Terrace in the Notting Hill area and we lived there until we left for the U.S. We expected to return to this after the year in California but big changes were in store.

Chapter 7
Palo Alto and the NASA Ames Research Center

Penelope and I flew to San Francisco on November 1, 1967. We were met by John Billingham and driven down to Palo Alto where we stayed in a very exotic motel. There was an approach avenue flanked by rows of Greco-Roman statues, the room itself had a purple carpet and furniture, and the walls were covered with mirrors. Welcome to California! We soon rented a house in Chabot Terrace which was not a salubrious part of Palo Alto but was just on the right side of the freeway. The other side took you into East Palo Alto, which was definitely on the wrong side of the tracks. I bought a used car and it was only 10 min drive down the freeway to the Ames Research Center in Moffett Field. The house was old but reasonably comfortable and there was a nice back garden with various fruit trees. I remember that at one stage we picked apricots and dried them in the sun as I had done in Australia. Penelope often drove me to work so that she could use the car during the day.

I soon found that the environment of the NASA Ames Research Center was very different from that of a university. First it was surprising that you were expected to arrive and leave at particular times rather like a business office. Although there were laboratories, there was much less intellectual curiosity and scientific discussion than in an academic setting, and many of the people behaved more like managers than researchers. In fact one of the complaints of the Center was that there were too many Chiefs and too few Indians. There were also a lot of petty regulations. For example I remember that when I left at the end of the year, I had to check in with a large number of chiefs of sections to make sure that I wasn't removing anything that they needed. In fact I had not been in most of these sections and certainly had never met the chiefs.

In spite of this it was a reasonably productive year. This was the period when computers were first being used extensively in the kind of work I was interested in, and I spent much of my time learning to use the language, Fortran. I had actually been on an introductory computer course in London but the equipment there was very primitive and all the information was fed into the machine using punched tape.

J.B. West, *Breathing on the Roof of the World*, Springer Biographies,
DOI 10.1007/978-1-4939-7122-0_7

The Ames Research Center had excellent computing facilities for that time because much of the work at Moffett Field was concerned with aerodynamics.[1] For example they had several state-of-the-art wind tunnels. I spent a good deal of my time developing a computer model of the lung using distributions of ventilation and blood flow to elucidate the effects of ventilation-perfusion inequality on pulmonary gas exchange. Kelman in the U.K. and Olszowka and Farhi in the U.S. had shown how it was possible to represent the oxygen and carbon dioxide dissociation curves with computer subroutines, and this meant that it was possible to solve complicated equations numerically for the first time. I like to think that one of the papers I wrote at Ames was a classic (Fig. 7.1). Certainly when this work was continued a year later at the University of California, San Diego, it paved the way for the multiple inert gas elimination technique which has revolutionized methods of measuring distributions of ventilation-perfusion ratios in patients with lung disease.

Although I was excited by my work on the computer analysis of pulmonary gas exchange I had an interesting setback. At one stage a blue ribbon Presidential Science Advisory Committee visited our life sciences group, and I briefly presented the work that I was doing. One of the members of the group turned out to be Wallace Fenn, the former chief of Hermann Rahn, and one of the grand old men of pulmonary physiology. When I told him about the work he seemed to be completely unimpressed and unaware of its potential. I like to think that this was one of the few occasions when he was mistaken.

Even though the computing facilities at Ames were excellent there were many frustrations in those early days. The information to be fed into the machine was in a deck of punched cards and frequently batches of these would be left to run during the night. It was often maddening to find that a single error on one card stopped the whole program in its stride and consequently the night's run was wasted.

I also carried out an experimental project with the help of John Maloney who had joined me from Hammersmith. It was a somewhat complicated experiment but basically we were looking at the uptake of radioactive xenon by the blood in a saline-filled lung. We were able to show that proximal regions of the acinus (the unit of the lung where gas exchange occurs) were better perfused than the distal regions. Actually setting up this experiment showed some of the weaknesses of a government research laboratory. There were long delays in obtaining permission to use radioactive gases even though we had used them at much higher radioactivity levels at Hammersmith. The managers at Ames seemed to have little feel for the urgency of research, and one of their main objectives was to protect themselves if something went wrong.

[1]Moffett Field dates from 1930 when the land was acquired by the U.S. Navy. In 1931 an enormous hangar was built for the dirigible Macon but this crashed in the Pacific Ocean in 1935. In 1939 the National Advisory Committee for Aeronautics (NACA) established the Ames Research Center as a base for aeronautical research, and when NASA was formed in 1958 it became the NASA Ames Research Center. The facilities now include extensive life sciences laboratories and 14 wind tunnels including apparently the world's largest.

Respiration Physiology (1969) **7**, 88–110; *North-Holland Publishing Company, Amsterdam*

VENTILATION-PERFUSION INEQUALITY AND OVERALL GAS EXCHANGE IN COMPUTER MODELS OF THE LUNG

JOHN B. WEST[1,2]

Biotechnology Division, NASA, Ames Research Center
Moffet Field, California 94035, U.S.A.

Abstract. The effects of increasing ventilation-perfusion inequality on overall gas exchange were studied in digital computer models of the lung. Ventilation/unit volume and perfusion/unit volume were distributed log normally with respect to lung volume. Ventilation-perfusion inequality was found to affect the transfer of carbon dioxide nearly as much as oxygen. The reasons for the misconception that inequality does not interfere with carbon dioxide transfer are discussed.

The effects of changing overall ventilation, blood flow, inspired oxygen, hemoglobin concentration, and the acid-base status of the blood were investigated when oxygen uptake and carbon dioxide output were held constant at normal values. In general, mismatch of ventilation and blood flow in a lung caused the arterial oxygen tension to fall and the carbon dioxide to rise; increasing overall ventilation rapidly restored blood carbon dioxide tensions to normal but improved oxygen tensions little. Considerable improvement in oxygen tensions occurred, however, if the oxygen dissociation curve was made linear. In spite of large increases in inspired oxygen, severe hypoxemia and large alveolar-arterial oxygen differences remained when the inequality was severe. The results help to explain how the lung maintains oxygen and carbon dioxide transfer in the face of ventilation-perfusion ratio inequality and may be useful in characterizing the degree of ventilation-perfusion inequality in real situations.

Alveolar – arterial P_{O_2} difference	Perfusion/volume ratio
Alveolar gas composition	Pulmonary gas exchanges
Alveolar ventilation	Ventilation/perfusion ratio
Dead space	Ventilation/volume ratio

Although the general principles underlying the relations between ventilation, blood flow and gas exchange were understood by early investigators (for example, HALDANE 1922), quantitative analysis only became possible following the work of RAHN (1949) and his colleagues, and RILEY and COURNAND (1949). Since then considerable progress

Accepted for publication 25 January 1969.

[1] Work done during the tenure of a National Academy of Sciences Senior Postdoctoral Fellowship.

[2] Present address: Department of Medicine, University of California at San Diego, La Jolla, California.

88

Fig. 7.1 My paper on computer analysis of pulmonary gas exchange that resulted from work at the NASA Ames Research Center. This was used for the subsequent development of the multiple inert gas elimination technique which has proved to be so valuable in understanding ventilation-perfusion inequality in patients with lung disease

It was disappointing to find that the Ames Research Center had almost no contact with the exciting developments in manned space flight. These were the days when the Apollo program was in full swing but unfortunately all that work was done at the NASA Johnson Space Center near Houston, Texas and it could have been on Mars for all the contact we had. I gradually realized that there were tremendous jealousies between the various NASA centers and that turf was protected at all cost.

I hoped that Penelope enjoyed her period in Palo Alto. Of course, initially we did not know anybody initially except the Billinghams and Maloneys but gradually she made some friends and indeed at one luncheon she met Nancy Reagan, the wife of the then Governor of California. We did have some memorable skiing weekends. We would drive up on the Friday evening to the Lake Tahoe area, stay in one of the resorts and drive back on the Sunday. It must have been only a 3 or 4-h drive across one of the bridges over San Francisco Bay and then across the Central Valley into the Sierras. For some reason we did not see as much of San Francisco as we should have. The only freeway from Palo Alto at that time was old 101 and that was generally crowded. However we were less enterprising than we should have been in exploring that fine city.

While I was at Ames I talked to John Billingham about the possibility of making measurements of lung function on the astronauts some day. Of course these were early times in manned space flight and, for example, Neil Armstrong did not land on the moon until a year or so later. However John encouraged me to put together an application and this was sent off to NASA headquarters shortly before I left the Center. In it I proposed to measure various aspects of lung function before, during, and after spaceflight using relatively simple and well-proven tests based around a mass spectrometer. Wonderful to relate, the application was successful and when I arrived at UCSD on April 1, 1969 a check for $100,000 plonked down on my desk (figuratively speaking). As can be imagined, this made a big impression on Eugene Braunwald, the Chairman of the Department of Medicine who had recruited me. Our NASA support continued uninterrupted from that day for over 35 years! The program that we carried out mainly on the Spacelabs in the Shuttle but later on the International Space Station is described in Chap. 10. This followed the same basic pattern as that set out in that original application. This whole program has been enormously stimulating and is certainly some justification for the decision to spend a year at the Ames Research Center.

However the year there had another very unexpected dividend. I knew Bert Fung, Ben Zweifach and others in the bioengineering group at UCSD from their work on the biophysics of blood flow, and I made arrangements to visit them fairly early on during my stay at Ames. Nothing particular came of that. However in June of 1968 I attended one of the annual pulmonary conferences held in Aspen, Colorado. There I saw my old friend Averill Liebow who used to be the Chairman of the Department of Pathology at Yale University Medical School. He was an anglophile and I had met him at meetings in London. He had recently moved to become head of the Department of Pathology at the new UCSD Medical School although the students had not yet arrived. The first year started classes in the fall of 1968. I well remember having morning coffee with a group of people outside the

Fig. 7.2 Eugene Braunwald (1929–) who was the founding Chairman of the Medicine at UCSD Medical School and who recruited me to UCSD. One of the most prominent academic cardiologists in the world, he left UCSD in 1972 to become Hersey Distinguished Professor of Theory and Practice of Physic, Harvard Medical School

lecture hall when Liebow came up to me and asked me whether I knew about the new medical school in San Diego. I had never heard of it and told him so. He asked me whether I would have any interest in visiting it and perhaps becoming part of it. Of course I was very noncommittal, but he arranged for Eugene Braunwald who was the Chairman of Medicine to invite me down to La Jolla to give a seminar (Fig. 7.2).

I spoke about our work on the effects of gravity on the lung and this was well received. Then it turned out that Braunwald was looking for somebody at the interface between physiology and medicine to help set up the physiology program. The UCSD Medical School was unusual, perhaps unique at the time, in that the plan was to teach physiology to medical students out of the Department of Medicine. One of the reasons for this was that the departments of biology and biochemistry on the main campus where the medical school would be situated were so strong. The founders of the medical school argued that the basic preclinical science teaching could be done very well by them. In fact physiology was, and still is, represented by a Division within the Department of Medicine, not by a separate department. Braunwald indicated that he would like to pursue the possibility of my coming to UCSD, and he asked me where my wife was. When I told him he said have her fly down to San Diego and we can show her around. This was good American entrepreneurship!

Now Penelope had been reading the San Francisco Chronicle in Palo Alto and the columnist Herb Caen had a great loathing of Southern California. In fact we knew people who claimed that they would not even fly over Los Angeles! However Penelope took the next available plane down, we were shown around, and I think she immediately realized that La Jolla would be a wonderful place to live in. The upshot of all this was that Braunwald gave me an offer that you couldn't refuse, and after some delay I received a formal letter inviting me to join the faculty of UCSD Medical School as a Professor of Medicine step 2, a very attractive opportunity indeed.[2] So perhaps this is another justification for my decision to go to Ames.

A very important event during 1968 was the birth of our son, Robert, on September 21. There are some odd features to this story. Penelope had her prenatal care at the hospital of Stanford University. All went well although there was one unfortunate incident when she was skiing. She collided with another skier and was briefly knocked out but rapidly recovered and there were no untoward consequences for Robert as far as we could tell. However when the time came for his birth, the obstetrician was missing because he was attending the annual Stanford-San Jose football game! Even more worrying was that Robert was a breech presentation, and even I could remember from my obstetrics lectures that this was certainly to be avoided in a primipara (woman who had not previously given birth). Happily in spite of this, Robert grew up to do his undergraduate work at Brown University, and medical course at Washington University in St. Louis, where he also obtained a Ph.D. in molecular biology. He is now a full professor in the Pathology Department at Stanford and very productive. However I always think that Stanford owes him something.

Talking of Stanford I had quite a bit to do with the medical school during our year in Palo Alto. There were links between the school and the Ames Research Center, and I got to know a number of the faculty members there quite well. One was Herb Hultgren, a prominent cardiologist who had contributed much to high altitude medicine. He was the first person to describe high altitude pulmonary edema in the English literature, and he was also the first person to suggest that hypoxic pulmonary vasoconstriction might be uneven. This later turned to be a critical step in our explanation for the mechanism of high altitude pulmonary edema, which is briefly discussed in Chap. 11. During my period in Palo Alto, Stanford was looking for a new Chief of the Pulmonary Division and I gave them a seminar. Although negotiations never went as far as a formal offer I think this might have been arranged. However in the event I believe that going to the new medical school at UCSD was the correct decision.

In November 1968 we returned to London to sort things out before going to San Diego. We had to sell the house in Denbigh Terrace, and I also bought a Triumph TR6 sports car to export to California. During this time we lived with Penny's parents in Beech Coppice and I commuted each day to Hammersmith to tidy things up. It was an unusually wet winter and seemed to rain every day which made it easier to leave London. Of course I had some regrets in leaving Hammersmith

[2]The University of California was founded in 1868 and presently comprises 10 campuses including five medical schools. Three of these at Davis, Irvine and San Diego were started in the 1960s. At the present time the University has 250,000 students, 190,000 faculty and staff, and has produced 62 Nobel Laureates.

where I had spent 15 years off and on. I think Penelope's parents were also disappointed because the original arrangement was for us to go to California for 1 year only, and their daughter was now going to be some 5000 miles away. This was all the more difficult because their other daughter, Joan, lived in New York City. I was a bit apprehensive about starting a new research program from scratch, and it was not entirely clear what my responsibilities would be as Chief of the new Physiology Division in the Department of Medicine. However many other people had made similar moves, and at the age of 41 I had more of my professional life ahead of me than behind me.

Chapter 8
University of California, San Diego, 1969–1981

We flew from London to San Diego on March 31, 1969. It was a very long flight with young Robert of just over 6 months, and because the plane was delayed there was no one to meet us. I remember that it seemed to take forever to get from the airport to La Jolla because the freeway was closed for some reason. We arrived at the Andrea Villa motel and stayed there in one room for a couple of weeks.

However we quickly began to become organized. Eugene Braunwald (Fig. 7.2) had arranged a University car for me which was very thoughtful. A remarkable event had occurred a few weeks earlier when we were in London. We received a letter from Marge Bradner, the wife of one of the faculty members at UCSD, congratulating me on my appointment and asking whether there was anything she could do to help us find a house. The letter referred to all sorts of goodies such as an ocean view, swimming pool, nearness to the campus, etc., and it made a big impression on us both. We had been used to the generally unhelpful attitudes of real estate agents in London and to get a letter like this weeks before we arrived seemed extraordinarily hospitable. Of course Marge Bradner had an ulterior motive which was to sell us a house if possible, but this did not detract from her very helpful attitude. Marge and her husband Brad, a physicist, became good friends of ours. They had both worked at Los Alamos during the war when the nuclear bomb was developed. The upshot of Marge's letter was that Penelope was driven around by her for a week or so to look at available houses near the campus and we soon had a good feel for what was possible. We chose a house in Greentree Lane just across the road from the campus in the La Jolla Farms area (Fig. 8.1).

One of our problems was that we had little money. My salary at Hammersmith had always been small, and although I made much more during my fellowship at the Ames Research Center, it turned out that we were not allowed to take much currency out of the U.K. because of export restrictions. In fact I remember writing to the Governor of the Bank of England one day asking for an exception, which I think was given. Oddly enough, this restriction actually had some advantages in the long run. Because we were not allowed to take much cash out of England, we

J.B. West, *Breathing on the Roof of the World*, Springer Biographies,
DOI 10.1007/978-1-4939-7122-0_8

Fig. 8.1 Back garden of our house in Greentree Lane. Shortly after we arrived I was astonished to learn that we were only 10 min or so from the Torrey Pines Glider Port which is one of the best-known sites in the world for flying radio-controlled gliders

bought some nice antique furniture and had this shipped across. This has given us a great deal of pleasure and I hope will do the same for our children.

Of course I needed a mortgage to buy the house that cost $70,000. Fortunately my father was able to come up with some $30,000 that he sent from Australia and this was sufficient for the mortgage. This was the first time that my family had sent me any money apart from the air ticket I referred to earlier, but it was a critical piece of help. My employment at UCSD officially started on April 1 and my salary then increased considerably. I was not used to an income like this and went to see the local bank manager to ask him what priced house he thought I could afford. He seemed very surprised by this question.

It took a few weeks for the owner of the house that we were buying to move out, and of course we did not want to stay in a motel all that time. Marge Bradner very kindly organized a temporary place for us. It was a brand new house that had been bought by another new faculty member, presumably through Marge Bradner, and we kept it warm for 2 or 3 weeks. It was in the new La Jolla Shores Heights development and I remember it well. It had a very high ceiling in the entrance way and a bright yellow wall-to-wall carpet.

In April 1969 UCSD Medical School was in its infancy. Although the first year medical class had started lectures in the fall of 1968, the School had few faculty members at this stage. Of course all the new appointments were senior professors, and a nice feature was that each one gave a lecture to a gathering of the entire faculty soon after they arrived. Another attraction was that the Dean of the Medical School, who lived only three or four houses from us, gave a party on his patio to welcome each new faculty member. The early faculty members of this new school included some stellar academics including Kurt Benirschke, Eugene Braunwald,

Louis Gluck, Elliott Lasser, Averill Liebow, Bob Livingstone, Arnold Mandell, William Nyhan, Kenneth Ryan, Edwin Seegmiller and Daniel Steinberg. There was a very lively intellectual atmosphere and a great esprit de corps. The School is now so large that I know relatively few of the faculty and even most of the people in the Department of Medicine are strangers to me!

At that time, none of the buildings on the upper campus where the center of the new Medical School was located had been completed. In fact I was the first faculty member to move into the Basic Science Building, which was the first building to be finished. In the meantime, I had a small room down at the University Hospital in Hillcrest some 10 miles away. Incidentally when I visited the School in the fall of 1968 to interview other faculty members before I was recruited, I found that most of them were housed in the Scripps Institution of Oceanography in La Jolla Shores because there was no space on the upper campus.

When I eventually moved into the Basic Science Building I had a small office with room for a secretary next door, and one large laboratory room that of course was absolutely bare. Setting up from scratch was a daunting prospect. However I soon arranged to bring over two technicians from my lab in London on a temporary basis, George Kingaby and Rosalind Stevenson. They had worked extensively with me on the isolated perfused animal lung preparation that had proved to be so productive (Fig. 6.2) and I set this up at UCSD. They stayed about a year and then returned to London. One of the first large purchases on the NASA grant was a respiratory mass spectrometer so that we could continue our work on pulmonary gas exchange.

A very fortunate event occurred in the fall of 1969. Peter Wagner had graduated in medicine from the University of Sydney and spent some time doing research with John Reid whom I knew from a period that he spent at Hammersmith with me in the 1950s. John was a brilliant young doctor who tragically took his own life a few years later apparently because of frustrations in developing the new Westmead hospital. After a year with John Reid, Peter decided that he should come to the U.S. for a period. He wrote to Hermann Rahn but, fortunately for me, Hermann said that his department was full at that time. I think it was Hermann who recommended that Peter write to me. He did so but it turned out that there was another Dr. West at UCSD who had recently moved to Taiwan, and many of the letters addressed to me went to him in error. Fortunately he used to send them back, and one of these was from Peter Wagner. When I received his letter enquiring about postgraduate work, I was attending the fall meeting of the American Physiological Society at the University of California, Davis. I remember that I immediately telephoned Peter from that campus with great difficulty, and told him that we would be happy to have him. This was one of the luckiest events in my early days at UCSD because Peter has turned out to be a brilliant academician, and he took over the Physiology Division a few years ago when I thought it was time for me to step down.

One of the new obligations that I had at UCSD Medical School was to teach physiology to first year medical students. Of course the teaching at the Postgraduate Medical School in London was limited to doctors who had already obtained their medical degrees. I found this new teaching responsibility very stimulating and enjoyable. The annual ritual of presenting the basics of the whole area of respiratory physiology has something of a religious experience about it. It has something in common with monastic communities who read the whole of the Scriptures on a 1-year cycle. Although teaching first year medical students might sound easy because the material is so basic, it can be very challenging. In fact many senior medical people find it difficult to teach at this level because they cannot develop the mindset that is needed. Also first year medical students can ask very penetrating questions. One of the stories I tell is that early on in the course, you sometimes get questions that are very original and that are difficult to answer. However after a few weeks the students have developed the mindset of their teachers and unfortunately most of these innovative questions disappear.

I mentioned earlier that the best book for first year medical students when I started was Julius Comroe's *The Lung,* and indeed this was the book that persuaded me to pursue a career in respiratory physiology. However I soon became dissatisfied with the book in the context of teaching first year medical students, partly because it put undue emphasis on the control of ventilation. This was Comroe's primary interest, but is not a critical area for medical students to understand. Far more important is the essential function of the lung which is gas exchange. Therefore, early on I prepared some lecture notes that better suited the orientation that I thought was most valuable for the students, and after a couple of years I decided to assemble these notes into a book, *Respiratory Physiology—The Essentials.* This was published by Williams & Wilkins (Fig. 8.2). This book soon became very successful, and has remained the most popular book in its area for medical students and for many other related students as well. It has been translated into 15 languages and I cannot resist showing the translations in Russian, Japanese, Farsi and Chinese (Fig. 8.3). The last two were pirated, that is unauthorized with no royalties. It is the only book of the 24 that I have authored or edited that has made any money worth mentioning, but it has done very well. In fact when the C.P.A. who prepares my annual tax return saw the royalties after a year or so, he suggested that I form a corporation so that some of the royalties could be put in a tax-deferred pension fund. This was very valuable advice.

Of course in academic medicine in the U.S., a critical responsibility is developing outside funding. Many people do not realize that the financial support from the Medical School itself is minimal. Specifically the state paid only a portion of my salary, and a small part of the salary of a secretary. It is true that it also provides the space for my office and laboratory, and it pays for the utilities, but that is the end of it. As the Division of Physiology prospered and the teaching responsibilities increased, for example in the graduate program in physiology, two other state-funded positions were added, and parts of the salaries from these two people also came from the state. However any academician worth his salt has to generate substantial amounts of outside support.

My NASA funding which began on the day I started at UCSD was the main source for the first couple of years. However I then submitted a proposal for a Training Grant from the National Institutes of Health so that we could have three or four postdoctoral trainees and one or two graduate students. Then as our research

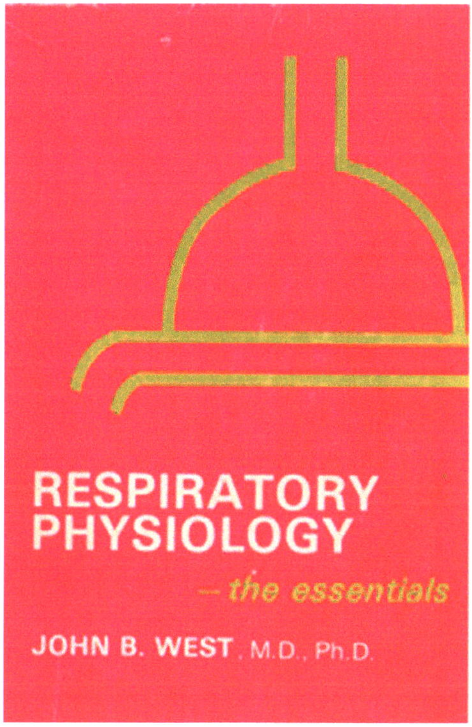

Fig. 8.2 First edition of *Respiratory Physiology—The Essentials* which was published in 1974. It has remained the most popular book in its area for medical students

program expanded beyond the area that could reasonably be supported by NASA, I prepared an application for an N.I.H. Program Project grant. All of these were successful and continued uninterrupted for 30 or so years. These responsibilities were taken over by Peter Wagner when I stepped down as Chief of the Division a few years ago.

My philosophy from the beginning was that a relatively small but focused unit could be very effective, and in many respects was more satisfying than a big empire which needed a great deal of administration. Partly as a result of this thinking, the Division of Physiology has remained relatively small with three state-funded positions and a few other faculty members on "soft" money, that is they are supported from outside grants. Since Peter Wagner took over as Division Head, it has expanded somewhat although it is still relatively small as divisions in the Department of Medicine go with relatively modest funding and a modest amount of space. In fact the last became a serious problem especially when the NASA program expanded to handle the development of hardware for the Spacelab experiments. The space on the UCSD campus was then completely inadequate and we were forced to rent space in the adjacent General Atomics facility.

Looking back on the last 40 years I sometimes wonder whether this initial philosophy was correct. To take one example, the Division of Pharmacology started off in the same way as a division of the Department Medicine, but Palmer Taylor

Fig. 8.3 Four translations of *Respiratory Physiology—The Essentials* in Russian, Japanese, Farsi and Chinese. The last two were unauthorized

who became chief after Steve Mayer stepped down was adamant that the Division must become a Department to prosper. He managed to carry this through although it was an enormous undertaking going all the way up to the University President's office. More recently he became the main force behind developing the new School of Pharmacy and Pharmaceutical Sciences at UCSD. Perhaps I should have taken a course more like this, although the more limited aims of our Division have always kept the amount of administration to a minimum. The result is that I have always had plenty of time for research in addition to the teaching. The Division has developed a substantial international reputation for its research.

I like to think that my career blossomed during my first few years at UCSD. I became Chairman of the Medical School faculty in 1972–1973, and I was a member of one of the N.I.H. Study Sections in the early seventies and its Chairman from 1973–1975. I was also a member of the Physiology Test Committee of the National Board of Medical Examiners in the early 1970s. The role of this group is to compose questions for the various state and other exams. Incidentally there is an anecdote about this. After 5 years at the UCSD Medical School I was required to take the California exam for medical licensing. Until then I had had a complimentary license as a foreign V.I.P. The licensing exam was tough for me because several of the subjects, for example molecular biology and immunology, did not exist when I was at medical school. However I did find two or three of my own questions on the physiology portion of the exam!

An unexpected honor was the Ernst Jung Prize for Medicine, which was presented to me in Hamburg, Germany in 1977. This was an impressive occasion, which of course Penny also attended. I gave a brief presentation, and received a chunky gold medal and the considerable sum of 50,000 DM. I am afraid that one of the reasons why I was chosen for this award was that I was friendly with one of the Swiss pulmonary physicians on the awards committee. However this did not diminish my pleasure in receiving the award and each year we are invited to be present at the prize-giving ceremony.

On the family side the big event in 1971 was the birth of our daughter, Joanna on January 30. By then Penelope and I had begun to consider moving out of the small house in Greentree Lane although this was certainly very comfortable. We had very interesting neighbors by the name of Suess. Hans was an eminent geochemist at the Scripps Institution of Oceanography. His grandfather, Eduard Suess, had been very famous indeed, and had written a classical monograph in German translated as *The Face of the Earth*. Hans himself worked amongst other things on the carbon-14 method of dating old materials, and he had helped to calibrate this by measuring the amount of the isotope in the rings of the very old Bristlecone Pine trees from the White-Inyo region of California. For this he used a very low level counting facility on the top of Mount Soledad in La Jolla. Ruth, his wife, was very artistic and their house was exquisitely decorated. She was very lively and one of her anecdotes was that Hans would have received a Nobel Prize if she had been willing to sleep with someone or other. While the Suess' were away for a period, the house was occupied by the prominent neuroscientist, Stephen Kuffler, who worked at the Australian National University, but was spending some time at UCSD. One of my many missed opportunities was to get to know him during this period.

If you visit the region around the UCSD campus today, it is very difficult to imagine that in 1969 when we moved here the whole area was essentially undeveloped. I can remember sitting in my office in the Basic Science Building and looking east where there was hardly a building to be seen. Most of the inhabitants of the area were rabbits. In particular the La Jolla Farms area between

Fig. 8.4 The back garden at 9626 Blackgold Road when we bought the house in 1972

the campus and the ocean was mainly vacant lots, and we learned that a developer was putting up some houses. We bought one of these on $1\frac{1}{3}$ acres of land and I sometimes say that it was one of the smartest financial moves that we have ever made. Certainly we would not be able to afford any house there now. The house was poorly constructed and, for example, the insulation on the flat roof was minimal so that it was boiling hot in summer and freezing cold in winter. However over the years we have made many changes and the house is now extremely comfortable. We still have photographs showing the garden in its virgin state of a broad expanse of clay (Fig. 8.4). A short distance down the road were some riding stables, and one of my memories is seeing horses walking across our newly-sown back lawn in the morning on their way down to the beach. In fact, the area had been developed by William Black as a Beach and Bridle Club. Black's impressive adobe house is now the University House where the Chancellor lives.

Moving into the new house was a considerable challenge but we were young and enjoyed it. The area available for the garden was enormous. Indeed we had the impression that many people were reluctant to buy such a large property because it entailed so much work, but coming from London where space was at a premium (the tiny back garden of our house in Denbigh Terrace was an example) we loved the idea of the space. Penelope is fond of gardening, and with a little help from our

Fig. 8.5 The back garden of the house now as a result of the efforts of Penelope and helpers

friends the result has been very rewarding (Fig. 8.5). Over the past few years the La Jolla Farms area has become very desirable with some enormous mansions being created. In one sense this is a pity because the area has lost the neighborhood feel that it originally had. When we moved in there were a number of university faculty in the area, but there are few who can afford to live here now.

Chapter 9
American Medical Research Expedition to Everest, 1981

Of all the major events in my professional life, this expedition stands out as the most remarkable. Even now I find it difficult to believe that I led an expedition to Mt. Everest that was successful in putting five people on the summit, and that we obtained the first physiological measurements in that unique environment.

As indicated earlier, I had no particular interest in high altitude physiology prior to the Silver Hut expedition of 1960–1961, although of course I knew a little about the topic because it is closely related to respiratory physiology. However the great success of that expedition, including the fact that we made extensive measurements at the very high altitude of 5800 m (19,000 ft), and particularly the studies that I made with Mike Ward on the Makalu Col at the extreme altitude of 7440 m (24,400 ft) meant that I became well-known in the area of high altitude physiology. There were a number of publications from the expedition and I was invited to give many talks.

One of the fascinating findings from the Silver Hut Expedition is shown in Fig. 9.1, which plots the maximal oxygen consumption (a measure of the amount of work that the body can do) against the barometric pressure that falls as the altitude increases. The left lower point is the one that Mike Ward and I obtained on the Makalu Col. The prediction from these data was that essentially all the oxygen available on the summit of Mount Everest would be needed for the basal oxygen requirements of the body, for example providing enough oxygen for the heart to pump and the brain to work. In other words, it looked as though there would be no oxygen left over to enable someone to climb to the summit. Thus these measurements suggested that the mountain would never be climbed unless people took bottled oxygen with them.

After the Silver Hut Expedition I often wondered whether it would be possible to return to the Himalayas for further physiological studies, and in particular whether it might conceivably be possible to get measurements high on Everest, including perhaps on the summit itself. The first time that I discussed this with anybody was in September 1974 when I met Jim Milledge during a visit to London and we talked about the possibility of making simple measurements of lung and heart function at

J.B. West, *Breathing on the Roof of the World*, Springer Biographies,
DOI 10.1007/978-1-4939-7122-0_9

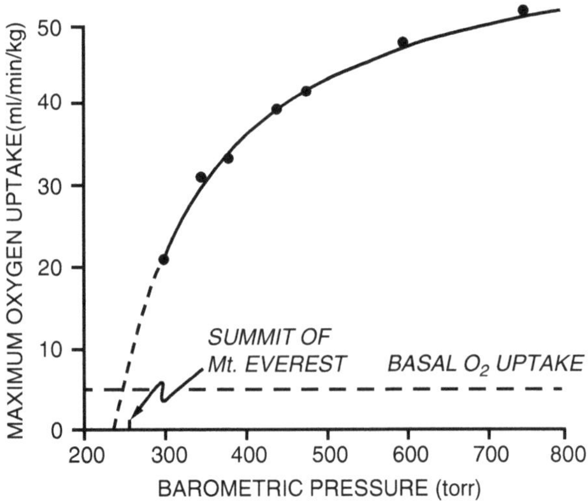

Fig. 9.1 Maximal oxygen consumption in acclimatized subjects plotted against barometric pressure which falls as the altitude increases. The point on the far left was obtained by Mike Ward and myself on the Makalu Col, altitude 7440 m. Note that extrapolation of the line to the expected pressure on the summit of Mt. Everest predicts that all the available oxygen will be needed for the basal oxygen requirements of the body

extreme altitude on Mount Everest. Jim had been contacted by Arnold Heine, a New Zealander, about an expedition to Everest that was planned for 1977, and Jim and I discussed the possibility of adding a small scientific group to this expedition. This prompted me to apply to the National Heart Lung and Blood Institute in early 1976 for a contract to make a plan for human physiology measurements above 7500 m, and the contract was actually funded partly through the support of Dr. Claude Lenfant, an old friend of mine who had become Director of the Lung Division of the Institute. Fortunately he had a strong interest in high altitude physiology. I looked at small portable instruments that could be carried by a climber including the "Medilog," a miniature slow-running tape recorder that could record four channels of data over 24 h. This new device had been developed by Dr. Norman Holter for monitoring electrocardiograms during daily activity, and I knew him slightly because he had a home in La Jolla.

While we were studying this equipment I met Dr. Chris Chandler who was preparing for an expedition to Everest called the 1976 American Bicentennial Expedition. He agreed to take the Medilog to the summit and succeeded in obtaining good data on the ECG up to about 7300 m. Above that a broken wire precluded further recording but this test certainly suggested that it was feasible to obtain data at extreme altitudes.

In the event the proposed collaboration with the New Zealand expedition did not eventuate, and this was not surprising because trying to combine scientific work with an expedition whose primary objective is to reach the summit results in all sorts of tensions. There is no doubt in my mind that if the aim is to make rigorous physiological measurements at extreme altitude, this has to be done by a dedicated scientific expedition.

At about this time, I met Dr. F. Duane Blume who was in the Department of Biology of the California State University at Bakersfield. The link was that he had developed an oxygen system for climbers that was of interest to the New Zealand expedition. Duane had been a member of the International Everest Expedition in the spring of 1971. He had trained at UC Berkeley with Nello Pace, a well-known high altitude physiologist, and had worked in the UC White Mountain Research Station near Bishop. Duane was a good friend of John Evans who had been a climbing leader on the 1971 expedition, and who had shown remarkable poise in the face of extremely difficult circumstances including the death of a climber who, it was alleged, had been abandoned by other expedition members. I met John Evans and was enormously impressed by him, and the upshot was that the three of us decided to try to organize the American Medical Research Expedition to Everest (AMREE) with the principal objective of obtaining the first physiological measurements on the Everest summit.

Here it should be pointed out that almost no measurements had previously been made above those of ours at 7440 m altitude during the Silver Hut expedition some 15 years previously except for a few alveolar gas samples. Even today no additional physiological measurements have been made on the Everest summit beyond those done during AMREE, and so our plans in the seventies were extremely ambitious. Nevertheless, in January 1978, the three of us sent an official application to Nepal for a climbing permit and this was granted in June 1978 for the post-monsoon (Fall) period of 1981. While we were waiting for the permit, a remarkable event occurred in May 1978. This was the first ascent of Everest without supplementary oxygen by Rheinhold Messner and Peter Habeler. This proved that it could be done, but only heightened the interest in what physiological mechanisms made it possible.

The letter of permission from Nepal is something of a collector's item. The erratic behavior of the Nepalese typewriter contrasted greatly with the ponderous diplomatic sentiments: "His Majesty's Government of Nepal presents their compliments to the Embassy of the United States and begs to inform them that permission has been granted for the medical research expedition to Everest for the post-monsoon period of 1981." However of course this was wonderful news and we could now begin planning in earnest.

Duane, John and myself met shortly after permission for the expedition was granted and we formed a California corporation with ourselves as the three directors. The main advantage of this was that we could then apply for non-profit status that meant that money donated to the expedition was tax-deductible. We also thought that the corporation might limit our liability if some disaster occurred. I then sat down and composed a 50-page grant application to the National Institutes of Health to raise the seed money. I requested $157,721 for 3 years beginning in July 1979. Although the fall of 1981 seemed a long way away, the time soon disappeared when you started counting backwards.

When the application got to the Study Section which assigns its scientific priority, it was given a tough time. Although most of the members were enthusiastic about the research, a minority of three voted disapproval (Fig. 9.2). One of the chief reasons was that they thought the project was too dangerous, and that government funds should not be used for such a hazardous enterprise. There were also some technical questions about the statistical significance of results on so few people. The latter issue did not bother me too much. One look at the back of the Moon was

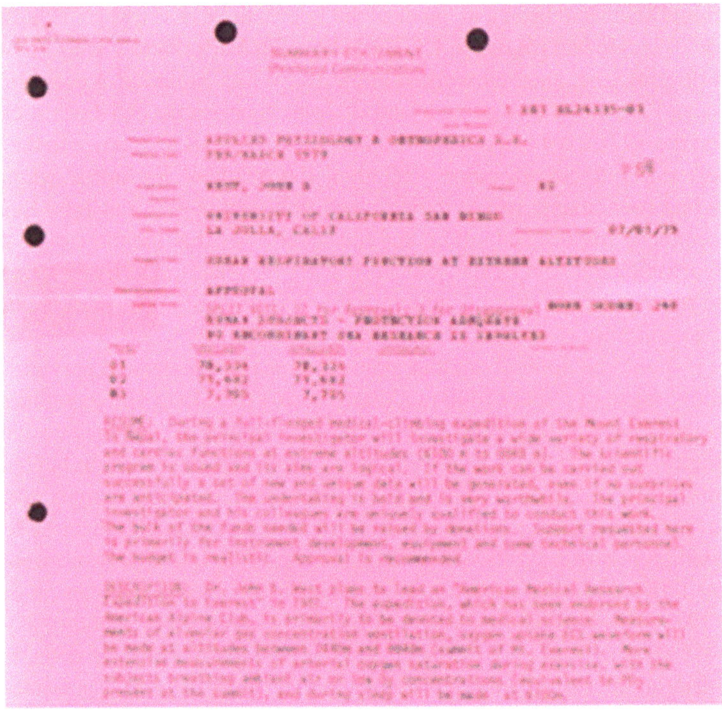

Fig. 9.2 First part of the "*pink sheets*" showing the report of the NIH Study Section on the application to support the scientific work of AMREE. Note that 3 of the 15 members of the Study Section voted for disapproval, reducing the score to 248 (100 is highest, 500 is lowest). Nevertheless the application was just funded

enough to show that it was not made of green cheese. A vote of disapproval from several members of a Study Section is usually a mortal blow and so I was pessimistic. However a miracle occurred in 1979 when the Institute reported that just enough funds were available for the application to get through. This was critically important and was pivotal as far as developing the scientific program was concerned. In fact the timing was such that if the application had not been funded at this stage, we would have had to call off the whole expedition. Incidentally N.I.H. funding has since become much tighter, and certainly an application with the same priority score as ours would not now be funded. This was one of a series of lucky events that kept occurring throughout the story of this expedition.

With money in hand it was now time to choose members of the expedition. Duane, John and I decided that we needed three kinds of people. First it was necessary to have very experienced climbers who had shown their ability to go above 8000 m and who could put in the route to the summit. Next we decided to recruit some "climbing scientists." These were all very strong climbers but they all had M.D. degrees as well. Initially the thought was that these people would instrument the climbers who were going to the summit, but in the event there was

Fig. 9.3 Members of the expedition at Kathmandu prior to the trek in. Back row (*left-to-right*): Evans, Boyer, Blume, Kopczynski, Lahiri, West, Jones, Hackett, Graber, Winslow, Samaja, Schoene, Pizzo. Front row: Sarnquist, Korich, Peters, Weis, Porzak. Absent: Lowe, Maret, Milledge

very little difference between the climber and climbing scientists on the mountain itself. In fact one of the ironies was that of the three Americans who reached the summit, two of them were climbing scientists. A third group was chosen to run the laboratories at the Base Camp and at Camp 2 where many scientific measurements would be made. For the record, the climbers were John Evans (deputy leader in charge of climbing), David Jones, Chris Kopczynski, Jeff Lowe, Glen Porzak and Michael Weiss. The six climbing scientists were Steve Boyer, David Graber, Peter Hackett, Chris Pizzo, Frank Sarnquist and Brownie Schoene. The eight remaining scientists were Duane Blume (deputy leader in charge of logistics and finance), Sukhamay Lahiri, Karl Maret, Jim Milledge, Rick Peters, Michele Samaja, Bob Winslow and myself (Fig. 9.3).

This is an appropriate place to clarify the first paragraph of this chapter which refers to me as leader of the expedition. Some readers might have a mental image of a line of climbers ascending the final summit ridge with me at the front. Nothing could be further from the truth. All movements of the climbers on the mountain were planned by John Evans and I never ventured above Camp 2. In my book about the expedition[1] I likened my role to that of the Duke of Plaza Toro in Gilbert and Sullivan's opera The Gondoliers:

[1]Some of the scientific findings are described in West, J.B., *Everest – The Testing Place*, New York: McGraw-Hill, 1985. There is also an account in Chap. 11 of West, J.B., *High Life: A History of High Altitude Physiology and Medicine*, New York: Oxford University Press, 1998.

Fig. 9.4 Ed Hillary who agreed to be our patron, and myself, discussing some aspects of AMREE

In enterprise of martial kind when there was any fighting
He led his regiment form behind—he found it less exciting

I conceived the expedition but had no part in the logistics on the mountain.

Incidentally there were a number of similarities between AMREE and the Silver Hut expedition that took place 20 years before. Three of us (Jim Milledge, Larry Lahiri and myself) were left over from the Silver Hut Expedition, and the design of the Base Camp laboratory was similar to that of the Silver Hut in many respects. Furthermore, Ed Hillary who led the Silver Hut Expedition agreed to be the patron of AMREE and he also generously agreed to provide a foreword for the book that I wrote about the expedition (Fig. 9.4).

Of course the expedition members were carefully chosen by John, Duane and myself. The details are not relevant here but I cannot resist the temptation to describe the selection of Karl Maret. It was clear that someone with extensive bioengineering and technical experience would be essential if all the complicated equipment was going to work satisfactorily. Here providence intervened in a remarkable way. Someone from our UCSD laboratory was having dinner at a vegetarian restaurant in San Diego when he got into conversation with the waiter. He was taken aback to learn that this young man was a newly qualified M.D. who was waiting on tables to make ends meet. He suggested that Karl come to see me, and this was just after we had received the funds to develop equipment so I had my eyes open for someone trained in medical instrumentation with high altitude experience. However when Karl walked in my face fell. His curly hair hung down to his shoulders, there were beads around his neck, and he slopped in with sandaled feet.

Karl's story was that he had recently graduated with an M.D. from the University of Toronto but because he became disenchanted with traditional medicine, he did not do the required year of residency after graduation but instead joined a holistic medicine group. Their belief was that healthy living came about through meditation, vegetarian food, music, and dance. However the organization was in financial difficulties, and Karl was obliged to spend some of his time working as a waiter.

With my middle class background this did not seem like a promising beginning. However it turned out that Karl had earned a masters degree in bioengineering at Toronto prior to doing medicine, and furthermore that he had worked on a high altitude project carried out by the well-known high altitude physician, Charles Houston, on Mount Logan. Thus he actually had ideal qualifications and he eventually turned out to be one of the most valuable members of our expedition. He was responsible for much of the design of the miniaturized equipment and his contributions were crucial. However the two sides of Karl's personality were always evident. A discussion of some technical aspect of data recording could easily progress to a discussion of levitation.

A sketch of the mountain showing the expedition route is shown in Fig. 9.5, and the sites where physiological measurements were made is shown in Fig. 9.6. It can be seen that there were four sites on the mountain where we hoped to make measurements. The first was at the Base Camp (5400 m; 17,700 ft) where we placed a prefabricated hut that served as a laboratory. Just beyond the Base Camp is the start of the Khumbu Ice Fall, a frozen waterfall of ice about 700 m (2300 ft) high, which is a daunting obstacle for expeditions to the Everest summit from the Nepalese side. Above the Ice Fall is the high valley known as the Western Cwm, so named by Mallory in 1921. Then comes the southwest face of Everest, and our highest camp

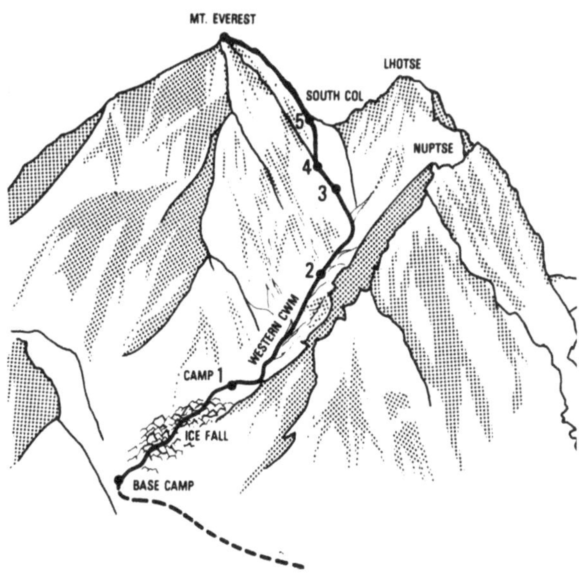

Fig. 9.5 Sketch of Mt. Everest from the southwest showing the route taken by the expedition

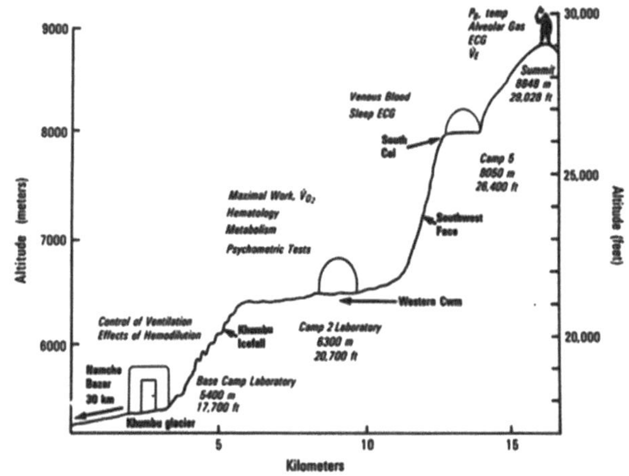

Fig. 9.6 The four sites on the mountain where medical research experiments were carried out

was just above the South Col at an altitude of 8050 m (26,400 ft). Most ambitious of all was our hope to obtain some measurements on the summit at 8848 m (29,028 ft). I say ambitious because when I looked back at the five Everest expeditions prior to ours, I found that not one had reached the summit. Success depends very much on the weather. If this is bad, forget it. Another important factor is whether enough climbers can tolerate the extreme altitude and remain healthy enough.

Before our departure from the U.S. we were flattered to receive four official documents wishing us well. One was a letter from the White House with a fine embossed letterhead in which President Reagan wrote that "The goals and aspirations of your mission capture the imagination of our people and symbolize the drive for accomplishment that is the heart of the American spirit" (Fig. 9.7). There was also a supportive letter from the Governor of California, a proclamation from the Governor of Colorado, and a flowery resolution from the California Legislature.

The members of the expedition arrived in Kathmandu, the capital of Nepal, at the beginning of August and we spent most of that month walking into the Base Camp. Here we set up the prefabricated laboratory consisting of foam-insulated plastic panels that were bolted together (Fig. 9.8). In the original design of the expedition we had planned to put this laboratory hut at Camp 2, but an early severe storm that paralyzed the expedition for several days convinced us that it required too many loads to carry through the Khumbu Ice Fall.

The Nepalese regulations about the timing of Everest expeditions were strict, and we were not permitted to start in the Ice Fall until September 1. However at 4 a.m. that morning, several of the climbers and climbing scientists were up and working on the route and good progress was made. Unfortunately a few days later there was a fierce storm that prevented us from going into the Ice Fall for 6 days with the result that it was not until September 15 that the main laboratory camp, Camp 2 at 6300 m (20,700 ft), was established. The laboratory structure there was a Hansen Weatherport tent consisting of an aluminum frame covered with fiberglass blankets (Fig. 9.9). It was kept warm with a propane heater and supplied with

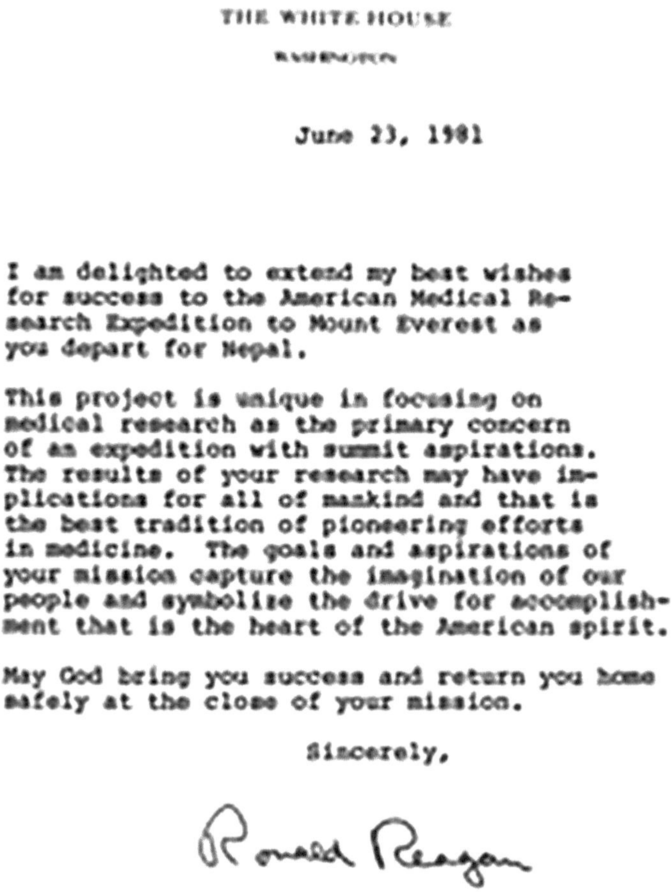

THE WHITE HOUSE

WASHINGTON

June 23, 1981

I am delighted to extend my best wishes for success to the American Medical Research Expedition to Mount Everest as you depart for Nepal.

This project is unique in focusing on medical research as the primary concern of an expedition with summit aspirations. The results of your research may have implications for all of mankind and that is the best tradition of pioneering efforts in medicine. The goals and aspirations of your mission capture the imagination of our people and symbolize the drive for accomplishment that is the heart of the American spirit.

May God bring you success and return you home safely at the close of your mission.

Sincerely,

Ronald Reagan

Fig. 9.7 Letter from the White House in which President Reagan wished the expedition success

electrical power from gasoline generators and solar panels. An extensive research program was carried out here.

The highest camp, Camp 5 at 8050 m (26,400 ft), was not established until October 11 when Chris Pizzo measured the barometric pressure there, and took a

Fig. 9.8 The prefabricated hut that served as the laboratory at Base Camp (5400 m; 17,700 ft)

Fig. 9.9 The Weatherport tent that was our laboratory at Camp 2 (6300 m; 20,700 ft)

series of alveolar gas samples on Davey Jones, Kop and himself. However high winds and very cold temperatures forced the group to return to Camp 2, and a second group of climbers on October 15 who reached Camp 5 were also battered back by the fierce winds with an extreme chill factor. In fact at this time many of us despaired of the expedition reaching the summit because the weather had been so severe.

Fortunately on October 21 the weather improved and Chris Kopczynski and Sherpa Sungdare left Camp 5 at 7 a.m. and reached the summit at 11:30 a.m., both using supplementary oxygen. Chris was keen to make physiological measurements on the summit and had been well trained to do so but unfortunately he could not find the appropriate equipment at Camp 5 because it had been buried by snow and ice as a result of the terrific storms. He therefore had to go to the summit without the equipment and consequently could not make any physiological measurements there. By this time there were only two or three climbers, or climbing scientists, left who were still fit enough to make a summit attempt.

However on October 23 Peter Hackett, Chris Pizzo and two Sherpas, Nuru Jangbo and Young Tenzing, reached Camp 5 and prepared the science equipment for the following day. Again the weather was favorable and Chris and Young Tenzing reached the summit at 12:30 p.m. Figure 9.10 is the most famous photograph from the expedition. It shows Chris Pizzo sitting on the summit of Mount Everest taking samples of alveolar gas from his lungs. He did this by rapidly exhaling as far as possible so that the last expired gas was trapped between two valves in the expiratory tube. He then pulled the lever shown, and this opened the valve of a pre-evacuated aluminum can, thus trapping a sample of the last expired gas. Six of these cans were arranged on a revolving fitting. Incidentally it was not easy for Chris to deliver the appropriate samples because of his extreme shortness of breath on the summit.

Incidentally there is an interesting story about the photo shown in Fig. 9.10. Of course it was taken by Sherpa Young Tenzing and as far as we know he had never taken a photograph before in his life. Chris related that the first thing that the Sherpa did when he was given the camera was to fiddle with some of the settings such as the focus and aperture. However as can be seen from the figure, the photograph was of exceptional quality with excellent focus and exposure, a very challenging feat in the extreme brightness of that setting!

Fig. 9.10 Dr. Chris Pizzo taking samples of alveolar gas from his lungs while sitting on the summit of Mt. Everest

Another important measurement that Chris made on the summit was the first direct measurement of barometric pressure. This pressure is critically important for understanding the physiology of extreme altitude because it determines the partial pressure of inspired oxygen, that is the amount of oxygen available to the climber in this extremely rarified air. Of course we had some indications of what the pressure might be from meteorological studies. However it was very satisfying to obtain the first direct measurement on the highest mountain in the world, and it turned out to be somewhat higher than expected at 253 mmHg, probably because the day was very fine and therefore the pressure was relatively high.

Chris also has his idiosyncrasies, and the first thing he did when he reached the summit was to take out a Frisbee and throw it into Tibet to get the world's high altitude Frisbee record (Fig. 9.11). This photograph also shows that Chris was using oxygen to climb to the summit. We gave the climbers the option of using oxygen if they wanted to, and although it could be argued that the summit measurements might have been "purer" if oxygen had not been used, the chances of getting to the

Fig. 9.11 Chris Pizzo about to throw a Frisbee into Tibet to make the high-altitude Frisbee record

summit without supplementary oxygen are extremely low, and I was glad that they chose to use it. This photograph also shows the turbine flow meter that recorded Chris's ventilation while he was climbing at an altitude 8300 m (27,200 ft) after he had temporarily turned off his oxygen. The data from the flow meter were recorded on a Medilog tape recorder and showed that his respiratory frequency was 86 breaths per minute! Since his stepping rate was probably about 10–15 steps per minute this means that there were about 6–9 breaths for every step! The Medilog also made a continuous record of Chris' ECG which was essentially normal.

The account so far omits one of the most extraordinary incidents of the whole expedition. When Chris and Young Tenzing set off for the summit on the morning of October 24, Chris could not find his ice axe. He had left it at Camp 5 a few days before but it had apparently been blown away or covered with snow and ice as a result of the terrific storms. However Chris was not going to lose the opportunity of a fine day to go to the summit. Instead he picked up a tent pole and started heading for the summit with that. Now there is no way that you can get to the summit of Mount Everest with a tent pole. The ice axe is a crucial piece of equipment and is essential for keeping your balance on the final summit ridge and for belaying your partner. It transpired that after Chris had been climbing for an hour or so, he was astonished to come across a splendid axe on the snow in front of him. It subsequently turned out to be the ice axe of a climber who had died some 2 years previously because after summiting she was not able to return to the highest camp before nightfall and she died of exposure. The ice axe had been rattling around on Everest all this time and presented itself to Chris just when he needed it. Of course it enabled him to get to the summit and, even more important from our point of view, it enabled him to obtain his extensive series of physiological measurements. This was an almost incredible stroke of good fortune.

Shortly after Chris' historic ascent and his collection of physiological data, another almost unbelievable event occurred. Peter Hackett and his Sherpa companion had set off for the summit a little later than Chris and Young Tenzing. However before they had climbed far, Sherpa Nuru Jangbo's feet became very cold, he thought he was developing frostbite, and he therefore turned around and went back to Camp 5. Of course Peter was bitterly disappointed. He did not have enough oxygen to get to the summit and back, and it was unthinkable that he would try to climb the final summit ridge without a companion to belay him. But he thought he would continue to climb slowly because he might be able to help Chris and Young Tenzing on the way down.

The three of them met at about 2 p.m. and although there is some disagreement about what was said, the upshot was that Peter decided to make a solo summit attempt! This was a very foolhardy decision because it was clearly too late in the day to reach the summit and return to Camp 5 in daylight. It should be emphasized that the final summit ridge is technically challenging with great exposure. One slip on the west side means a fall of about 2440 m (8000 ft), while if the climber goes through the cornice and falls on the other side, he drops rather further into Tibet. Foolish decisions are common at extreme altitude, presumably in part because of the cerebral hypoxia.

At any event Peter set out for the summit alone and Chris Pizzo waited for him at an altitude of about 8400 m (27,600 ft). Chris had a radio and was in contact with the rest of the expedition lower on the mountain so we could share his concern about Peter. The radio transmissions were recorded and transcripts are included in my book *Everest: The Testing Place* (see footnote 1). Our concern for Peter deepened when the sun began to sink in the west and Chris took a magnificent photograph of sunset on Makalu with Kangchenjunga in the far east of Nepal in the background (Fig. 9.12). Incidentally this photograph clearly shows the Makalu Col where Mike Ward and I made our cycle ergometer measurements back in 1961.

I have recounted the story of Peter many times but I still find it moving. He did indeed reach the summit and took a few photographs to prove that he had been there. However he was desperately tired which is not surprising because climbing alone on a dangerous section such as the summit ridge is very draining. He started to descend but when he reached a difficult pitch known as the Hillary Step (because it gave Hillary a lot of trouble during the first ascent in 1953) he encountered some soft snow and fell. Of course the normal result of this is a fall of about 8000 ft. to your certain death. However on this occasion Peter apparently fell some 8–10 ft. when one of his boots caught in a piece of outcropping rock and he found himself suspended upside down, completely alone, near the summit of Everest, with the evening coming on! Obviously a very dangerous situation. However he managed to

Fig. 9.12 Photograph taken by Chris Pizzo at an altitude of about 8400 m (27,600 ft) while he was waiting for Peter to return from his solo summit attempt. Makalu in the foreground shows the Col where Mike Ward and I made measurements of maximal oxygen consumption 20 years before. In the distance is Kangchenjunga

hang on to his ice axe and was flailing around with this when to his astonishment he exposed a piece of fixed rope which had been placed in this difficult pitch by some unknown expedition, presumably several years before. With the help of this rope he was able to extricate himself, and although he fell a second time, he eventually reached Chris who remained waiting for him. The two then descended in the dark to Camp 5 with the help of one headlamp. Once they got to Camp 5 they were safe because the two Sherpas were there and the climbers were able to re-hydrate and spend the night. On the following morning they took samples of venous blood from each other, and these were important in contributing to our understanding of the physiology on the summit. Everyone then descended and by October 31 the whole expedition was off the mountain.

All three American climbers who reached the summit were debriefed into tape recorders when they reached the safety of Camp 2. Transcriptions of the tapes make great reading and are included in *Everest: The Testing Place*. Chris also recorded a tape into a handheld miniature recorder on the way to the summit and this is preserved in the archival collection at UCSD. The main reason why he took the recorder was so that he could record the barometric pressure on the summit. We were certainly not going to rely on his memory! However there is a moving passage on the tape which he recorded while he was waiting for Peter. It goes as follows: "Now I'll continue to wait for Peter. I'm out of food and out of water, but I'm well clothed and feel I can make Camp 5 in an hour, but I'm not leaving here until dark. If I don't make it down and someone discovers this tape . . . it's very important . . .. please return it to Dr. John West, Department of Physiology, School of Medicine, University of California, San Diego, La Jolla, California."

What were the main scientific findings from the expedition? These were very extensive and some are too technical to go into details here. The information can be found in the numerous publications (see footnote 1). However perhaps the most interesting results concerned how lowlanders can tolerate the extreme oxygen deprivation of the Everest summit. It turns out that the most important physiological response is an enormous increase in the ventilation of the lung, that is the amount of air that is breathed. In fact on the summit Chris was moving some five times more air into his lungs per minute than at sea level. The result was that the partial pressure of oxygen in the lung was maintained at a viable although low level, and at the same time the pressure of carbon dioxide was reduced to almost unbelievably low levels. One of the consequences of this was that the blood becomes very alkalotic As a result, the oxygen affinity of hemoglobin is increased and this accelerates the loading of oxygen by the capillaries in the lung. Remarkably, mammals that live at high altitude such as the vicuna and llama and also have this increased oxygen affinity, and indeed if you look at a broader range of animals including reptiles and birds, they generally increase their oxygen affinity in a hypoxic environment. They use all sorts of biological strategies to do this. It is remarkable that the climate at extreme altitude does exactly the same thing but by a completely different strategy, that is by extreme alkalosis. Our measurements showed that on the summit, the alveolar PO2 was only about 35 mmHg, the PCO2 was between 7 and 8, and the pH exceeded 7.7. These numbers can be contrasted with the normal values at sea level that are respectively about 95, 40, and 7.4.

When I lecture about these findings to physicians who work in intensive care units, I often ask them whether they have ever seen a patient with such derangement of their normal blood gases as we saw on the summit, and so far no one has volunteered that he has. It is extraordinary that a climber on the Everest summit can tolerate such an amazing abnormality of his normal physiology and live to tell the tale.

A particularly interesting finding was the maximal oxygen consumption for a climber on the summit. As pointed out earlier in this chapter, previous measurements had predicted that a climber would have no oxygen available for physical work at this great altitude. Of course it was not feasible to take a cycle ergometer to the summit. However we got the same information by making measurements on climbers who were extremely well adapted to very high altitude at Camp 2, and giving them 14% oxygen instead of the normal 21%, to breathe. When we did this we found that the maximal oxygen consumption was just over 1 l per minute, which is about a fifth of the sea level value. However this extremely low level explains how it is just possible for some humans to reach the summit without supplementary oxygen. In fact when we went back to the accounts of Messner and Habeler who made the first ascent of Everest without supplementary oxygen, it turned out that their climbing rate fits well with our result.

Penelope had arranged a trek of wives and friends to meet the expedition as we walked out. The logistics were organized by Mountain Travel Nepal who did a great job. It was very impressive that Penelope was able to reach the Base Camp at about 17,700 ft., particularly as the last part of the route is on a glacier with difficult walking. She was duly impressed by the Spartan conditions at Base Camp. She had a little peripheral edema with puffiness of the eyes, but otherwise was in excellent form. There were several parties on the way out and the American ambassador to Nepal, Carl Kool, also put on a splendid party in Kathmandu. We had a wonderful reception at San Diego airport with Robert, Joanna and Joan, Penelope's sister, who had been looking after the children together with Penny's parents while she was away (Fig. 9.13).

AMREE had a major influence on my subsequent career. First, there were a number of scientific articles in major physiological journals. One title that I particularly like was "Pulmonary gas exchange on the summit of Mount Everest". There was a general article in the prestigious journal *Science*[2], and this was accompanied by a fine cover photograph shown in Fig. 9.14. The American Physio-

[2]West, J.B., Human physiology at extreme altitudes on Mount Everest, *Science* 223:784–788, 1984.

Fig. 9.13 Our return to San Diego airport with the box containing the alveolar gas samples. *Left-to-right*: Penelope's sister Joan, Penelope, Robert, Joanna, myself

logical Society organized a symposium on "High altitude and Man" and this resulted in a book edited by Sukhamay Lahiri and myself. I wrote a semi-popular book about the expedition titled "Everest: the testing place". One of my most ambitious books is "High Life: a history of high altitude physiology and medicine" which was published in 1998 and has become a standard work in the field. I see it cited by booksellers as a reference source from time to time. Another product was further editions of the textbook "High altitude medicine and physiology" edited with Jim Milledge and Michael Ward. This was started by Ward several years ago and has now become a popular reference text for people who need an answer to questions on high altitude medicine. The present edition is the 5th.

An important event for me was an invitation from the publishing company owned by Mary Ann Liebert to be the editor-in-chief of a new journal "High altitude medicine and biology". Initially I was reluctant to take this on because I thought that there were already plenty of journals. However Mary Ann was charismatic and persuasive, and she convinced me that there was a niche for a new journal in this area. Subsequent experience has confirmed this, and the Journal has become a leader in what is admittedly a rather narrow field.[3] I was editor-in-chief for 16 years, and then decided that it was time for new blood, and the Journal was taken over by Eric Swenson. The subscription base of the Journal is rather limited because the area is so specialized. However high altitude medicine is

[3]There is information about *High Altitude Medicine & Biology* on the web at www.liebertpub.com/ham/.

Fig. 9.14 Cover of the journal *Science* in which a general article on the scientific results of AMREE was published. The photograph was taken in the Western Cwm with Nuptse as the background

burgeoning partly because of the increasing number of people going to high altitude for pleasure, but also to work. In addition there is a large permanent high altitude population in the Andes and the Tibetan plateau that has been poorly served in the past.

There were also less tangible results of the expedition. I was elected a council member of the American Physiological Society, and in 1984–1985 became its president. This prestigious election owed much to my profile as leader of the expedition. I continue to have a strong interest in high altitude physiology and medicine and some of the later projects are discussed in Chap. 11.

श्री ५ को सरकार
पर्यटन मन्त्रालय
His Majesty's Government of Nepal
Ministry of Tourism

बिक्रम सम्बत २०३८ सालको शरद ऋतुमा संयुक्त राज्य अमेरिका देशका श्री डा. जोन बी. वेस्ट को नेतृत्वमा [illegible] पर्वतारोही दलले ८८४८ मिटर उच्चाई भएको सगरमाथा हिमचूली सफलता पूर्वक आरोहण गरेकोले यो प्रमाण–पत्र प्रदान गरिएको छ ।

This is to certify that Amer. Medical Research Exp. to Ever. expedition led by Mr. Dr. John B. West of U.S. of America successfully climbed 8,848 meter high Mt. Everest in the year 1981 (Autumn).

15 Nov. 1981
Kathmandu

Under Secretary
Mountaineering Section

Secretary

Fig. 9.15 Official document from the Nepalese Government confirming that our expedition had climbed Mt. Everest

As mentioned at the beginning of this chapter, I still sometimes find it difficult to believe that such a wonderful experience happened to me (Fig. 9.15). The expedition was very successful from a scientific point of view, and of course we were almost incredibly lucky with no serious injuries. I do not think that anyone deserves more than one experience like this in a lifetime.

Chapter 10
Studies of Astronauts During Spaceflight

As recounted in Chap. 7, I had submitted an application to NASA for a research project to look at pulmonary function of astronauts in space when I was at the NASA Ames Research Center in 1968. The reason for spending a year at a NASA facility was that we had been working extensively on the effects of gravity on the lung. Figure 10.1 summarizes some of the pulmonary changes caused by gravity. The apex of the lung has a low pulmonary blood flow compared with the base, and it was indeed our finding with radioactive oxygen 15 as discussed in Chap. 4 that opened up this whole area. We also showed that the upper regions of the lung had a reduced ventilation compared with the lower regions. The basic region reason for this is that the lung is very deformable, and it therefore distorts as a result of its weight. Therefore the alveoli at the apex of larger than those of the base just as a slinky spring distorts when it is held vertically. A related consequence is that the expanding pressure on the apical alveoli is greater than that on the basal alveoli. Milic-Emili and his colleagues in Montréal have also made important contributions to this area of research. The larger expanding pressure at the apex of the lung also means that the mechanical stresses affecting the structure of that region of the lung are greater. The bottom line is that the lung is exquisitely sensitive to gravity, and therefore we thought it would be valuable to know what happens in the weightlessness of space.

It was remarkable that the proposal that I submitted to NASA in 1968 was in fact funded. I say remarkable because this was still early in the US space program, and all efforts were focused on getting a man to the moon and returning him safely to earth. It was extraordinary that in this environment, NASA would provide money to look at pulmonary function. However as noted earlier, this was very fortunate for me because when I started work at UCSD on April 1, 1969, the first check arrived, and it turned out that we were continuously funded by NASA for over 30 years.

You have to be very patient to work on space projects. As will be seen, although funding began in 1969, the first experiment that we did on microgravity was not until 8 years later in 1977. This was looking at the effects of short periods of weightlessness obtained in aircraft during parabolic flight. The first experiments in

J.B. West, *Breathing on the Roof of the World*, Springer Biographies,
DOI 10.1007/978-1-4939-7122-0_10

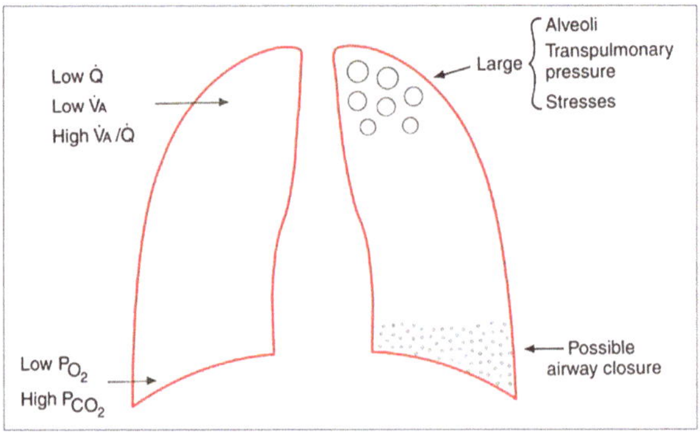

Fig. 10.1 Diagram summarizing the effects of gravity on the lung

Fig. 10.2 Learjet aircraft. This type of aircraft enabled us to make the first measurements of the effects of weightlessness on the lung

extended weightlessness did not occur until 1991, that is 22 years after the funding began. The last space experiment was in 1998, and unhappily shortly after that, the Shuttle program was discontinued. There is presently a low-key research program on the International Space Station.

The studies in aircraft flying a parabolic profile in 1977 produced very interesting results. The aircraft was a NASA Learjet operating out of the Ames Research Center (Fig. 10.2). In addition to the pilot, it accommodated one test subject, and David Michels, the physiologist conducting the test. The equipment was specially designed for this experiment and included a small respiratory mass spectrometer. To obtain a zero gravity environment, the pilot first put the aircraft into a climb, and then he moved the stick forward to push the nose down in a dive. This is known as a parabolic profile flight pattern. The G level was monitored by the pilot from a meter. It was possible to obtain zero gravity states of up to 27 s in this way although shorter periods were generally used. There is one advantage of a high performance aircraft over later measurements made in Spacelab in that it is possible to obtain other G levels such as 1 or 2 G as required. By contrast, Spacelab provides zero gravity all the time.

The experimental maneuvers were classical single breath nitrogen washouts that have been used for many years. Briefly the subject exhaled fully to residual volume, and then took a full breath of oxygen to total lung capacity, after which he made a vital capacity expiration. A bolus of argon gas was introduced at the start of the inspiration to give additional information.

The analysis of the data is a little complicated but briefly we examined the oscillations in the concentration of nitrogen and argon during expiration, and also the terminal rise in gas concentrations at the end of the breath. The oscillations are caused by uneven rates of emptying of the upper and lower regions of the lung because of the movement of the heart within the chest. The terminal rise is caused by closure of small airways at the base of the lung as a result of its distortion by gravity. Figure 10.3 shows two representative records. The upper one was obtained during level flight, and therefore normal gravity, and shows obvious oscillations in both the nitrogen and argon concentrations, and also a marked terminal rise for both gases. By contrast, the lower tracing that was obtained when both the inspiration and expiration were made in the absence of gravity, shows that the oscillations had almost disappeared, and also that the terminal rise was very much less. These tracings give information about the regional distribution of ventilation in the lung. Additional studies looking at oscillations in the expired carbon dioxide

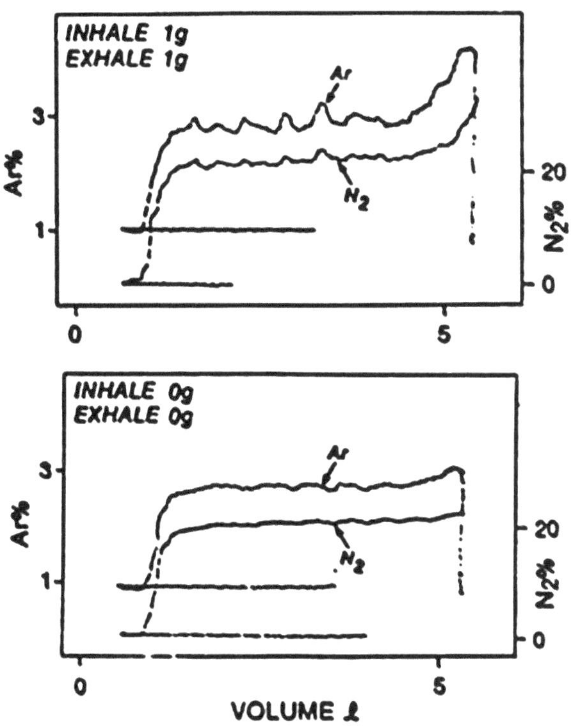

Fig. 10.3 Results from the Learjet experiments carried out at normal 1 G and at zero G. See text for details

tracings during a single expiration after a period of hyperventilation, gave information about the distribution of blood flow in the lung.

The results of these studies showed that virtually all the regional inequality of ventilation and blood flow seen in normal gravity conditions is abolished during short periods of microgravity. This was the first time that it had been possible to carry out studies like this in the absence of gravity, and the striking results made a big impression on physiologists and clinicians interested in this area. Another consequence of this work was to highlight the importance of further studies to be carried out on Spacelab.

The Spacelab SLS-1 program was a major scientific project involving a large team of investigators and an equally large group of support staff from NASA. In the event, the mission did not fly until 14 years after these preliminary measurements on the NASA Learjet. One of the reasons for the very long delay was the Space Shuttle Challenger disaster that occurred in January 1986. The tragedy occurred because the launch took place under unusually cold conditions and the O-ring seals of the solid rocket boosters were not designed for these temperatures. The result was that a little after a minute into the flight, one of the O-rings failed allowing hot gas to impinge on the external fuel tank with the result that it exploded and the crew of seven perished. This disaster triggered a series of new safety procedures for the Shuttle program resulting in a delay of over 2 years.

To be a principal investigator of one of the experiments on Spacelab SLS 1 and subsequent missions was certainly one of the highlights of my scientific career. First, the whole concept of setting up a laboratory in space on the Shuttle was very ingenious as Fig. 10.4 indicates. The laboratory was a cylindrical tank about 7 m long and 4 m in diameter, and it was located in the bay of the Shuttle where it could be accessed by crewmembers through a tunnel connected to the crew compartment. The interior of the laboratory was lined with instrument racks containing all the equipment (Fig. 10.5). The laboratory offered a "shirtsleeves" environment, in that the crew breathed 21% oxygen in nitrogen at normal barometric pressure, and the temperature and humidity were controlled. The result is that the only difference

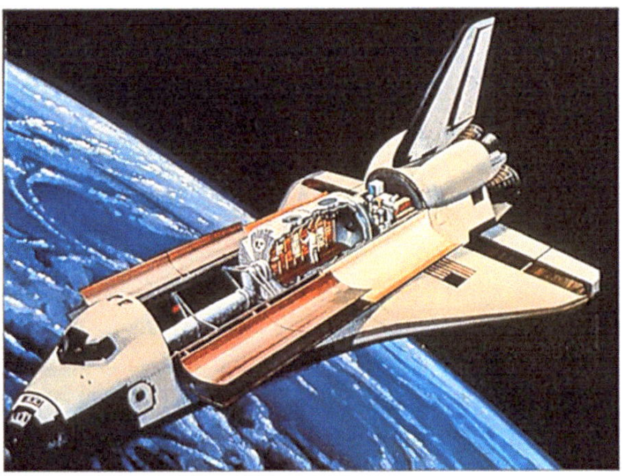

Fig. 10.4 Drawing showing the Spacelab in the bay of the Shuttle

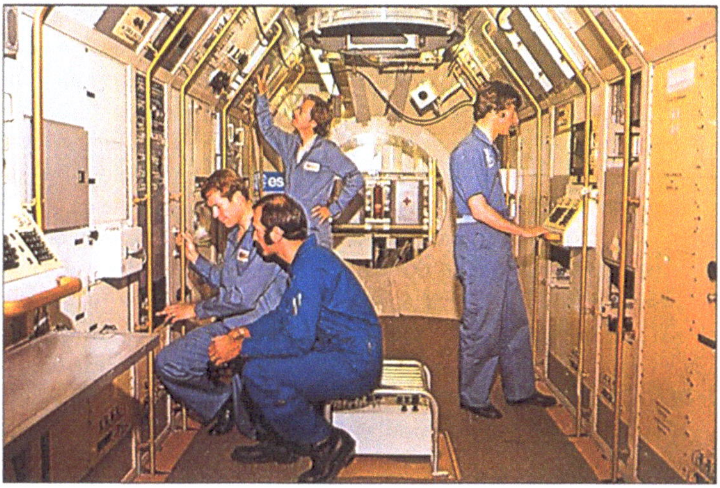

Fig. 10.5 Photograph of the interior of the Spacelab on the ground showing the instrument racks

in the environment compared with ground studies was the zero gravity, or as NASA likes to put it, microgravity.

There were some 10 investigations on human physiology in addition to our own including cardiovascular, renal/endocrine, blood, immune system, musculoskeletal, and neurovestibular studies. There were also projects on rats and jellyfish. All the principal investigators and their associates, together with the team from NASA staff, met on a number of occasions as the planning of the mission developed. Many of the meetings were at the NASA Johnson Research Center but others were elsewhere around the country. One of the main areas of discussion (and contention) was how much time could be allotted to each study. Crew time is at a premium during a flight like this. In fact prior to the mission, the NASA team put together a timeline that accounted for every minute of every person on board. Of course this was modified during the flight as various problems cropped up, and the timeline was reconfigured every evening for the subsequent days. The total Spacelab crew was made up of the commander and second officer, three career astronauts, and two mission specialists.[1] The latter were physiologists who trained for this particular mission.

I have been fortunate to have outstanding collaborators for the space experiments. The first person was David Michels who was largely responsible for designing and carrying out the studies of short-term microgravity on the Learjet. His place was later taken by Harold Guy who did much of the planning for Spacelab SLS-1. Finally Kim Prisk joined the group, and he has been responsible for the

[1]The crew members of SLS-1 were: Commander Bryan D O'Connor, Pilot Sidney M Gutierrez; Mission Specialists James P Bagian, Tamara E Jernigan, M Rhea Seddon; Payload specialists F Drew Gaffney, Millie Hughes-Fulford.

implementation of our studies on all four Spacelabs, SLS-1, SLS-2, LMS, and Neurolab. He is also led the work done on the International Space Station. Kim is extremely knowledgeable about all aspects of the space program and his contributions have been indispensable.

The package of tests that we designed for SLS-1 one was remarkably comprehensive given the relatively simple equipment that NASA would allow on board. There were nine parts to the package. First there were measurements of resting gas exchange where we measured the end-tidal PO2 and PCO2, oxygen consumption, and carbon dioxide elimination. A critical piece of equipment here was a respiratory mass spectrometer specifically designed for use in space. Next we obtained a measure of ventilation-perfusion inequality in the lung by measuring the changes in respiratory exchange ratio during a single full expiration. This was followed by a single breath nitrogen washout similar to the experiment that we carried out on the Learjet. In addition there was a similar test where we looked at the pattern of expired carbon dioxide following a period of hyperventilation. This had also been done in the Learjet and gave information about the regional distribution of perfusion. Next was a measure of the diffusing capacity of the lung using carbon monoxide. This is a standard test done in clinical pulmonary function laboratories. This was followed by a multi-breath nitrogen washout, which is another measure of uneven ventilation. At the end of that procedure, the oxygen concentration in the lung was very high and we exploited this fact to measure the diffusing capacity for carbon monoxide again. This allowed us to obtain information about its components. This was followed by a rebreathing measurement of total pulmonary blood flow (cardiac output) using the soluble gas, nitrous oxide, in the inspired gas. Finally there was a forced expiration test similar to that frequently done in hospital pulmonary function laboratories.

This package of 9 tests took only 31 min to perform and it was therefore a very efficient use of crew time. Furthermore we arranged the procedure so that the crewmember could perform the test on himself or herself by using a valve box with taps on both sides. Figure 10.6 shows Dr. Millie Hughes-Fulford in the process of doing the test during flight.

A very valuable feature of the experimental design was that there was a nearly continuous communication link between the spacecraft and the Johnson Space Center during the tests. This allowed us to see the data in real time on monitors and, if necessary, contact the crew about problems. Figure 10.7 shows Kim Prisk and myself watching the data as it came down. The upshot was that we had a comprehensive assessment of pulmonary function that took only about half an hour to obtain. Measurements were made on several occasions during the 9-day flight.

It would take too much space to go into details of the findings here. They are fully described in the resulting publications. However, as an example, Fig. 10.8 shows the results of measurements of the diffusing capacity for carbon monoxide. All the data are shown relative to the mean value for the standing posture before the start of the 9-day mission. It can be seen that the diffusing capacity increased sharply when the subjects were exposed to microgravity, and this increase was sustained over the 9 days. However on return to Earth there was a dramatic fall to the preflight value, and this was maintained for several days.

Fig. 10.6 Dr. Millie Hughes-Fulford carrying out our experiment named ALFE (Astronaut Lung Function Experiment) on Spacelab SLS-1

The reason for the increase in the diffusing capacity in microgravity is that blood that normally pools in the lower parts of the body because of gravity, redistributes in space. The result is that the volume of blood in the upper part of the body including the thorax substantially increases. This distends the pulmonary capillaries thus increasing both the available area for gas diffusion across the blood–gas barrier, and also the volume of blood in the capillaries. Both of these changes increase the diffusing capacity of the lung.

Here is a brief summary of what we learned about the effects of microgravity on the lung in exposures that lasted up to about 2 weeks. First, it was clear that there were substantial changes in the function of the lung during flight as would be expected from our previous work on the effects of gravity on the lung. We found that there was a more uniform distribution of ventilation and blood flow, although interestingly some residual inequality remained. As described above, the diffusing capacity of the lung was increased, and the cause of this was a redistribution of blood from the lower to the upper regions of the body. Incidentally this redistribution has other interesting effects. The resultant rise in the venous pressure in the

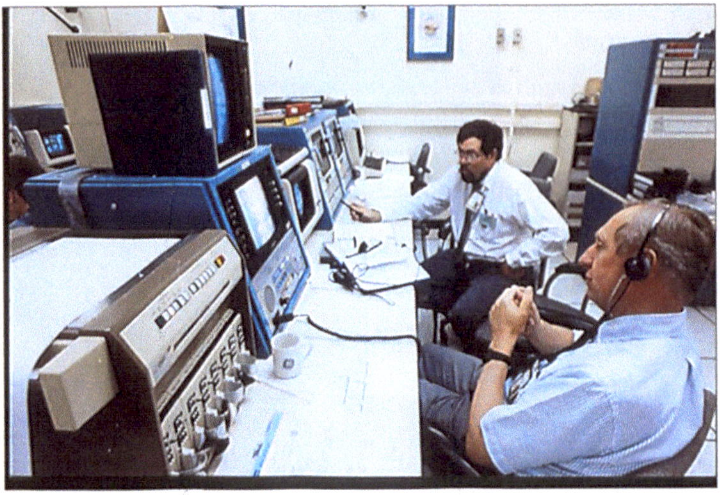

Fig. 10.7 The room at the NASA Johnson Space Center where the data came down in real time from Spacelab while one of our experiments was in progress. Kim Prisk is doing the work

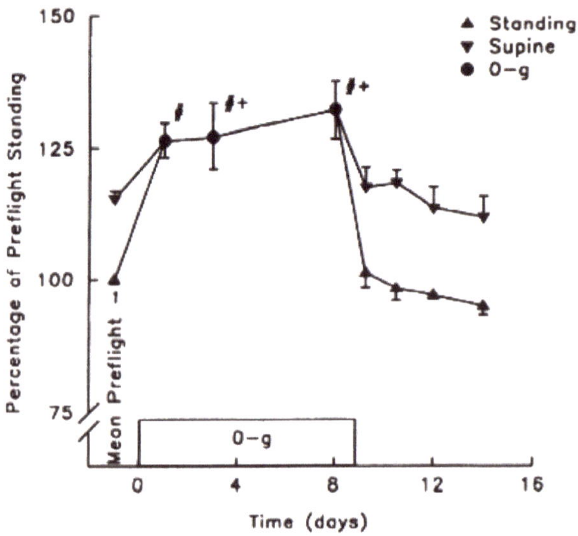

Fig. 10.8 Changes in the diffusing capacity of the lung for carbon monoxide before, during, and after the flight of SLS-1

upper parts of the body causes some mild interstitial edema that results in puffiness around the eyes. This can often be detected in photographs of astronauts in orbit. Some crewmembers develop hoarseness presumably due to mild edema of the larynx, and some complain of headache which may have the same basis.

Another finding was an increase in cardiac output early in the flight caused by a rise in cardiac stroke volume. There were also changes in the subdivisions of lung volume, for example the resting volume or functional residual capacity. The reason is that this is partly determined by the weight of the abdominal contents affecting the position of the diaphragm. We also found that there were alterations in the deposition of aerosol in microgravity although these measurements were actually made on aircraft in parabolic flights. This finding is of interest to NASA scientists who are concerned about the possible effects of inhaled toxic dust on the Moon and Mars. Most important was our conclusion that the overall gas exchange efficiency of the lung was in space was well maintained, with the result that astronauts can continue to live and work as on earth.

We also found some unexpected results, (incidentally this is always the case if careful measurements are made in a completely new environment). For example we looked at the behavior of two gases are very different molecular weights, helium and sulfur hexafluoride, when they were inhaled and exhaled. We found changes that apparently indicated differences in the rates of diffusion of the two gases. However diffusion rate depends on mass, not weight, and the results are therefore puzzling. Another unexpected finding was an apparent reduction in central venous pressure. It is difficult to reconcile this with the increase in cardiac stroke volume that we measured. The venous pressure measurements were made by another group of investigators.

The Spacelab flights had a maximum duration of only about 2 weeks and it is important to know what happens to pulmonary function with a longer exposure to microgravity. It was possible to get information on this when the International Space Station became available because crewmembers now typically spend much longer periods in this habitat. Kim Prisk was able to make measurements on astronauts before and after a 6-month period on the Space Station. He found that when measurements were made very soon after the crew returned to Earth, there were some changes compared with preflight. However pulmonary function returned to its pre-flight normal value within a few days.

The upshot is that although pulmonary function is clearly altered in microgravity, overall gas exchange continues to be satisfactory. Furthermore, after a long period in microgravity, pulmonary function returns to normal within a few days of return to normal G. This is important information for NASA scientists when they design long duration spaceflights, for example to Mars and back. The length of this mission is likely to be about 3 years. However the evidence to date is that pulmonary function is not likely to be a problem even in long-term space flight. There are other far more serious potential problems such as exposure to harmful radiation.

Many of our findings in microgravity were summarized in our book titled "Gravity and the Lung" with Kim Prisk as first author.[2] Other groups, especially in Europe, have now made a number of physiological measurements on the lung in space. However we had the satisfaction of doing this for the first time. It was a wonderful privilege to be presented with this unique opportunity.

---

[2]Prisk, G.K., Paiva, M., and West, J.B. (eds.), *Gravity and the Lung: Lessons from Microgravity*, New York: Dekker, 2001.

Chapter 11
Other Research Projects

When I look back over the years, five events particularly stand out.

1. Our early studies in the late 1950s using radioactive oxygen that showed for the first time the striking uneven distribution of blood flow in the human lung.
2. The Silver Hut Himalayan expedition in 1960–1961, which was my introduction to high altitude physiology.
3. Unraveling the physiological mechanisms responsible for the effects of gravity on pulmonary function at the Postgraduate Medical School in 1962–1967.
4. The American Medical Research Expedition to Everest when we made the first measurements of human physiology at the highest point earth.
5. The studies with the NASA Spacelab describing for the first time the effects of microgravity on the lung.

However I have been fortunate to be involved in a number of other interesting research projects and some of these will be briefly summarized here.

11.1 Stress Failure of Pulmonary Capillaries

This project started out with a puzzling clinical finding but ended up with new insights into a fundamental biological dilemma of the lung. The clinical puzzle was the mechanism of high altitude pulmonary edema (HAPE). This is a condition that affects some people when they travel to high altitude. It is characterized by marked shortness of breath, and in severe cases the patient coughs up pink frothy fluid. I related in Chap. 9 that one of the Sherpas on AMREE had the condition although I did not realize what it was at the time.

The recognition of the disease has an interesting history. Back in 1960, Herbert Hultgren who has been referred to previously, was visiting a hospital at high altitude in Peru when he saw a number of patients with pulmonary edema, and he recognized that this was a syndrome that he had not seen before. It had actually been reported in Spanish in the South American literature but was not known in

J.B. West, *Breathing on the Roof of the World*, Springer Biographies,
DOI 10.1007/978-1-4939-7122-0_11

North America. The mechanism of the edema was obscure. Initially left heart failure was suspected but measurements made during cardiac catheterization ruled this out. It was known that HAPE was associated with abnormally high pulmonary artery pressures. However these high pressures are believed to be caused by the constriction of small arteries upstream of the capillaries, and therefore there was no reason for the capillaries to leak fluid.

A breakthrough occurred when a group of research physicians including Peter Hackett and Brownie Schoene set up a clinic in a tent high on Mt Denali about 30 years ago to treat climbers with HAPE who were descending. The story goes that the physicians said "we would be happy to help you, but do you mind if we bronchoscope you first!". This was a very enterprising procedure in such a remote place, and incidentally it would be impossible to obtain permission for this procedure today. However these studies showed that the alveolar fluid had a very high protein concentration with many red cells, and clearly indicated that the walls of the pulmonary capillaries had been damaged. Hultgren had already suggested that the constriction of the small arteries might be uneven with the result that some capillaries could be exposed to the high pressure.

Strangely, nobody had apparently ever raised the pressure in pulmonary capillaries to abnormally high levels and looked up the ultrastructure of their walls. Fortunately we had Dr. Odile Mathieu Costello in our group at the time and she was an expert in electron microscopy of the lung. We therefore used an animal preparation where it was possible to raise the pressure in the capillaries, and study their ultrastructure. We were amazed to see that there were breaks in the capillary wall (Fig. 11.1). We concluded that the high stresses in the wall resulting from the increased pressure caused the structure of the wall to fail, and we introduced the term stress failure to describe this.[1] Stress failure is an engineering term referring to how a roof or other structure cracks or collapses when it is overloaded, for example, by very heavy snow.

The animal preparation allowed us to study many aspects of this phenomenon. One interesting finding was that if the pressure in the capillaries was first raised to a high level, and then the pressure was allowed to fall, and the tissue was fixed at the low pressure, many of the breaks disappeared. This fitted with the clinical observation that if patients with HAPE descend rapidly, they typically quickly improve. This is very unlike the situation with the pulmonary edema that occurs with other conditions such as heart failure.

No one had previously described ultrastructural damage in pulmonary capillaries exposed to a high pressure before. However when we looked through the literature, we found a number of instances where athletes at high levels of exercise developed edema, or coughed up small amounts of blood. This prompted Sue Hopkins in our lab to study a group of elite cyclists who rode up a hill outside the campus at top speed. They were bronchoscoped immediately after the exercise and fluid that was

[1]West, J.B., K. Tsukimoto, O. Mathieu-Costello and R. Prediletto. Stress failure in pulmonary capillaries. *J. Appl. Physiol.* 70: 1731-1742, 1991.

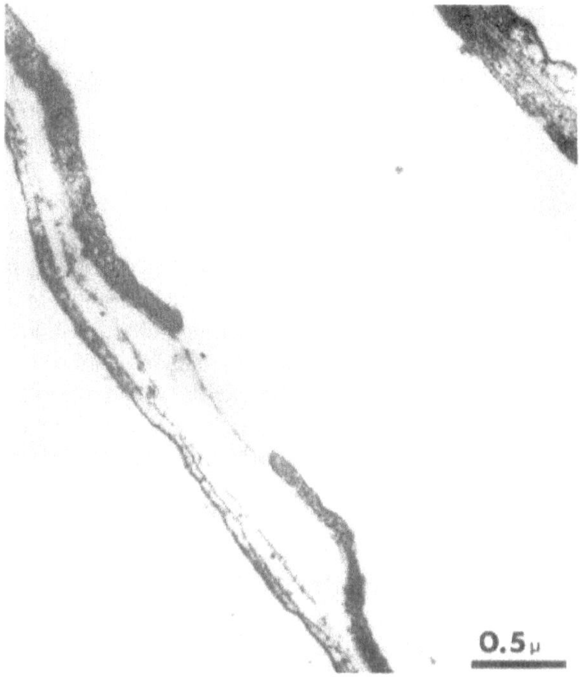

Fig. 11.1 Electron micrograph of a pulmonary capillary in an anesthetized rabbit showing breaks caused by an abnormally high capillary pressure. This is called stress failure

introduced into, and then removed, from the lung (so-called bronchoalveolar lavage). Increased protein levels and red cells were found in the fluid indicating that the alveolar walls had been damaged. She then studied another group of elite cyclists who pedaled for an hour but not at top speed. In this case no abnormalities were found in the lung fluid. The conclusion was that only under conditions of extreme exercise was the blood-gas barrier damaged.

One day when we were working on our animal preparation, somebody walked into the lab and told us that some racehorses bleed from their lungs after a race. I was not aware of this, but in fact it has been described from time to time for centuries. Various causes had been put forward over the years including moldy hay, but now we wondered whether the high levels of exercise in these animals could damage their capillaries. We therefore collaborated with colleagues at the University of California, Davis, and studied thoroughbred horses who were galloping on a treadmill at top speed. When we subsequently examined the lungs of some of these horses, we were able to see breaks in the pulmonary capillaries by electron microscopy.[2]

[2]West JB, Mathieu-Costello O, Jones JH, Birks EK, Logemann RB, Pascoe JR, Tyler WS Stress failure of pulmonary capillaries in racehorses with exercise-induced pulmonary hemorrhage J Appl Physiol 75:1097-109, 1985.

It turns out that the reason for the bleeding is that thoroughbreds have enormously high pressures in the pulmonary circulation during galloping at high speed. This is because of very high cardiac outputs required for racing. Remarkably, studies by others have now shown that all thoroughbred racehorses in training have hemosiderin-laden macrophages in their tracheal washings, and the implication is that they are all bleeding into their lungs.

There was an interesting sequel to this. I was invited to a meeting in Dubai where there were a large number of horse vetinarians and myself. I explained the mechanism of Exercise Induced Pulmonary Hemorrhage or EIPH as it is called, and of course the conference people wanted to know how to prevent it. All I could say was to have the horses run slower, but naturally this did not go down well.

We have now studied the phenomenon of stress failure of pulmonary capillaries in detail. The strength of the capillary wall is believed to be attributable to the presence of collagen IV within it, but the layer is extremely thin (about 50 nm, that is about 1000th that of a human hair). We believe that when the pressure in the capillaries rises too high, the hoop stress exceeds the breaking stress of the collagen and the capillaries bleed.

You might ask why do the capillaries have such thin walls if they break when the pressure rises greatly during severe exercise. The answer is that oxygen crosses the capillary wall by passive diffusion, and this means that the wall has to be extremely thin. The pulmonary capillary therefore has a dilemma. It needs to be extraordinarily thin to allow sufficient gas exchange to occur during heavy exercise. However at the same time it has to be extremely strong to withstand the very high stresses in its wall caused by the high pressures. This is a fundamental dilemma of the lung.

11.2 The Avian Lung

I cannot remember how I was first introduced to this topic but it has become a major area of our research. Probably we were prompted by observations made by my colleague, Frank Powell, on the avian pulmonary circulation when he was at UC Davis. Again this work owes a lot to the electron microscopy prowess of Odile Mathieu Costello, and later by Zhenxing Fu.

Some 300 million years ago, the ancestors of modern reptiles finally emerged completely from water and made a commitment to air breathing. Like modern reptiles they were cold-blooded and incapable of sustained levels of high physical activity. But from them developed the two great classes of vertebrates capable of high maximal oxygen consumptions, the mammals and the birds. A remarkable feature of these two groups is that although many their organ systems show considerable similarities, the lungs are radically different.

Figure 11.2 is a diagram comparing the general arrangement of the bird and mammalian lungs. One of the most important features of the bird lung is that it has separated the ventilation and gas exchange functions. This makes all the sense in

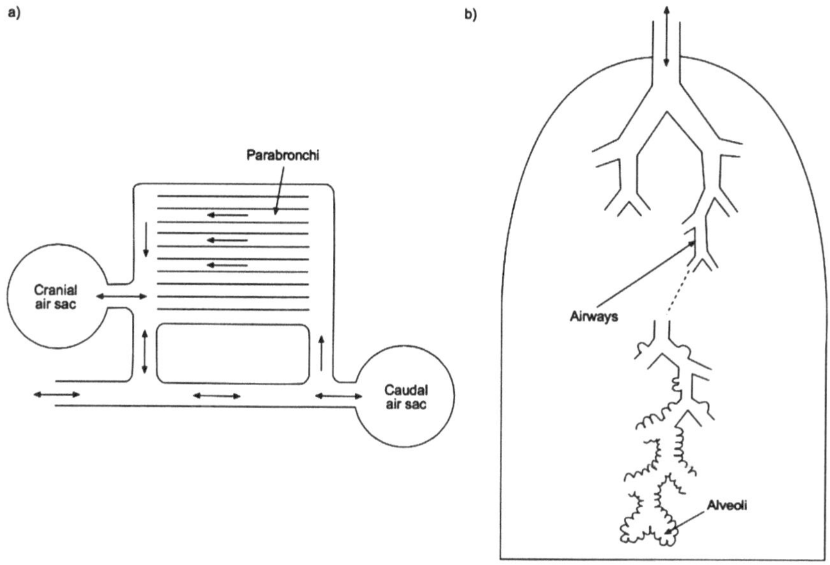

Fig. 11.2 Diagram showing the different anatomical arrangements in avian and mammalian lungs

the world. Why use delicate alveolar tissue to pump air when robust air sacs can do the same. In the bird, air is inspired into the air sacs, and from there it is pumped through the gas exchanging tissue called the parabronchi. This tissue is composed of blood capillaries, which are surrounded by air capillaries (Fig. 11.3) through which the gas flows, and the whole structure is essentially rigid. The pumping is done by the air sacs that take no part in gas exchange. By contrast, in the mammalian lung, the delicate gas exchanging tissue does the pumping, and it therefore undergoes changes in volume with each breath. One wonders whether this repetitive strain is a contributing factor in the common disease, emphysema, which is characterized by breakdown of alveolar walls.

Another feature that is obvious from Fig. 11.2 is that the bird lung uses a flow-through system for ventilation rather than the reciprocating pattern in the mammal. This latter has several disadvantages. One is that in the mammal, the alveolar gas to which the pulmonary capillaries are exposed necessarily has a lower PO_2 than the inspired air. By contrast, in the avian lung, the pulmonary capillaries are potentially exposed to the high PO_2 of the inspired air. Another drawback of the mammalian system is that because the inspired gas travels to the most distal alveoli partly by diffusion within the airways, uneven ventilation along the airway is always a potential problem, particularly during exercise when the oxygen consumption increases. A further disadvantage of the reciprocating system is that the terminal airways need to be relatively large so that the resistance to diffusion along them is low. By contrast, the terminal airways in the avian lung are extremely small. For example in the human lung, the alveoli have a diameter of about 300 μm, whereas

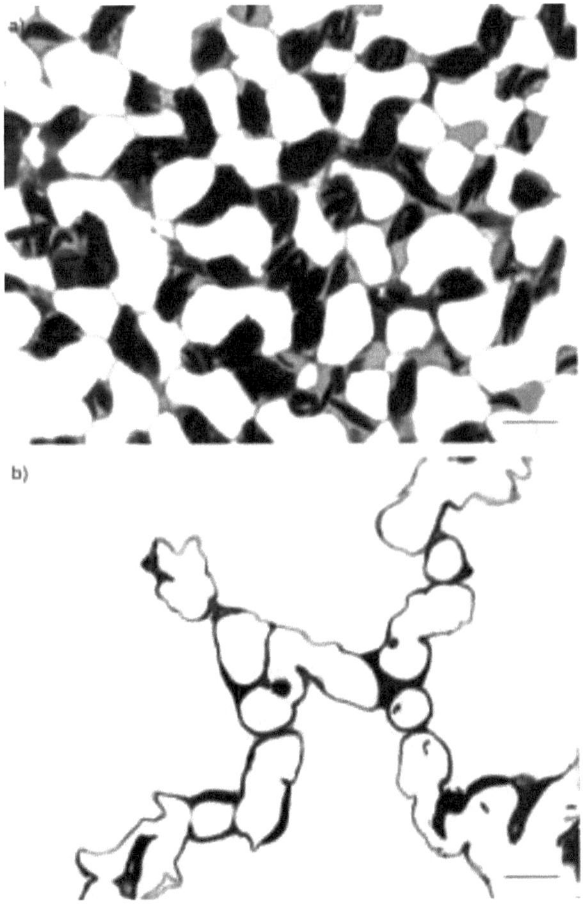

Fig. 11.3 Electron micrographs of pulmonary capillaries in avian (*bottom*) and mammalian (*top*) lungs

the air capillaries in the avian lung have a diameter of less than one 10th of this (Fig. 11.3).

The fact that the pulmonary capillaries in the bird are nested within a syncytium of air capillaries results in several additional advantages. Because the air capillaries contribute mechanical support to the blood capillaries, it turns out that the latter have thinner walls than in the mammalian lung. In addition, the walls are much more uniform in thickness. These differences are clearly advantageous for diffusive gas exchange.[3]

However there is one feature of the pulmonary capillaries in the bird that is puzzling at the present time. Because of the support offered by the surrounding air capillaries, the blood capillaries appear to be remarkably rigid, and not capable of

[3]West, J.B., R.R. Watson and Z. Fu. The human lung: did evolution get it wrong? *Eur. Resp. J.* 29: 11-17, 2007.

the distention that is typical of mammalian pulmonary capillaries. The distention and recruitment of the capillaries in the mammalian lung allows the pulmonary vascular resistance to fall markedly when the cardiac output increases as during exercise, and the result is that the rise in pulmonary artery pressure is small. This advantage does not appear to be present in the avian lung, but so far apparently no one has measured the pulmonary artery pressure in an exercising bird, for example during flight.

Together with Kirk Peterson who is an expert in cardiac catheterization of small animals, we designed an experiment to do this. He was able to place a small catheter in the pulmonary artery of anesthetized pigeons, and we could then continually measure the pressure by means of a device that was small and light enough to be carried by the bird during flight. Unfortunately technical and other problems proved too much. This is one of very few experiments that I have had to abandon.

These apparent advantages of the avian lung over the mammalian lung might raise the issue of why evolution took this route. However this is not a useful question. Evolution proceeds incrementally, and there is no designated endpoint. Instead random changes in the genome are selected for or against, and this is the explanation of how the two types of lungs developed. Nevertheless it is fascinating that the basic arrangement of the bird lung appears to be superior to that in the mammal.

11.3 The Elephant Lung

Some years ago, the American Thoracic Society had their annual meeting in San Diego, and they decided to have an outing to the San Diego Zoo which is well known. They asked me to give a talk on some aspect of comparative physiology, and I remembered reading that the elephant lung has a remarkable feature, namely that it is the only mammal without a pleural space. In all other mammals, this space between the lung and the chest wall allows the lung to slide over the inner surface of the wall. However in the elephant the space is obliterated by dense connective tissue. Interestingly, the fetal elephant has pleural space but this is replaced by connective tissue at about the time of birth.

This peculiarity of the elephant lung has been known for 300 years but nobody has come up with a satisfactory reason for it. Another unique feature of the elephant is its trunk, and I thought that it was likely that these two features were connected in some way. Of course, the trunk is used extensively for feeding, but it turns out that elephants also sometimes use it as a snorkel. Many naturalists and hunters in Africa have described seeing elephants cross shallow rivers by walking on the bottom, but breathing through the trunk that protrudes just above the surface of the water. In fact Aristotle described this behavior over 2000 years ago.

Snorkeling at depth creates very large pressure differences around the lung. These are shown in Fig. 11.4. If we assume that the bottom of the elephant lung is 2 m below the surface of the water, the water pressure at this depth is about

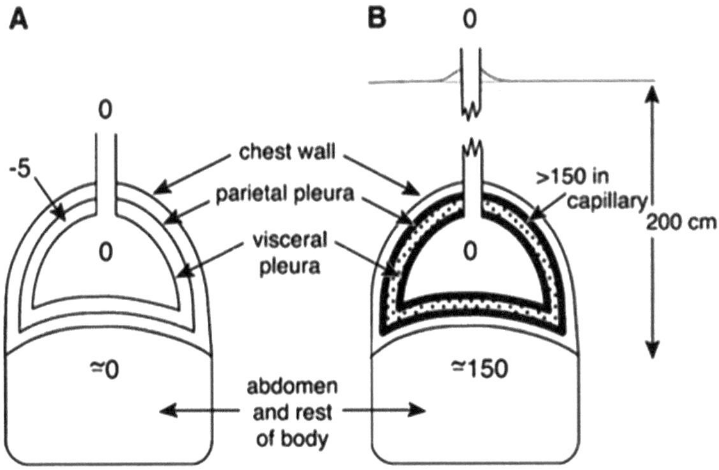

Fig. 11.4 Diagram of the pressures around the lung in a snorkeling elephant. The bottom of the lung is assumed to be 2 m below the surface of the water

150 mmHg, and all the vascular pressures in the animal are increased by this amount. However the pressure in the alveoli is close to atmospheric because the alveoli are connected to the surface by an open tube. Therefore at the outer surface of the lung, the pressure abruptly changes from 0 to 150 mmHg. The result is that the pressure difference across the walls of the capillaries supplying the pleura are so large that they would certainly break.

Evolution has provided a remarkable solution to this problem by replacing the normal delicate pleural membranes with sheets of dense connective tissue that prevent damage to the pleural capillaries. This seems to be a satisfactory explanation for this anatomic peculiarity in the elephant.[4]

Incidentally, there is evidence that the elephant has an aquatic ancestry, and it is possible that the trunk evolved to enable the animal to snorkel while living in water with obvious survival advantages. In particular, the elephant, (Order Proboscidea), and the sea cows (Sirenia) share a number of anatomical features.

The relative pressure changes shown in Fig. 11.4 also occur when the elephant raises water in its trunk for drinking or washing. Therefore this behavior, which is frequent in the elephant, is an additional evolutionary pressure for the unique anatomy of the pleural space. However raising water in the trunk takes only a few seconds whereas snorkeling lasts for many minutes and therefore seems to be much greater potential problem. On the other hand, it could be argued that the pressures during drinking constitute the major evolutionary pressure.

[4]West, J.B. Snorkel breathing in the elephant explains the unique anatomy of its pleura. *Resp. Physiol.* 126: 1-8, 2001.

11.4 Archive on High Altitude Medicine and Physiology

This is not a research activity in the usual sense, but it is a project that has given me much pleasure. Again I cannot recall exactly when it started although I think this was probably after AMREE when I had some material that I thought should be preserved. I therefore talked to Lynda Claassen who is the Director and Chief Curator of the Special Collections Library at UCSD, and was surprised to find that there already was a large archival collection. She suggested that we start a collection on high altitude medicine and physiology, and as far as I know this is unique in the world. It started in a small way, but has now grown to be substantial, and we have people from all over the world who visit it because of some special interest that they have.[5]

One of the major parts of the collection dates from when Griffith Pugh died in 1994. As Chap. 5 relates, he was the physiologist on the first ascent of Everest in 1953, and he organized the physiological component of the Silver Hut expedition in 1960 of which I was a member. I knew Griff well and was aware that he kept meticulous records and had a large collection of research material. When I was in London, I contacted his widow whom I knew, and asked her what was going to happen to his papers. Oh, she said, they are out in the barn and we plan to burn them. I replied don't do that; donate them to the University of California, and we shall keep them in our archive. She agreed, and shortly after this I visited her home just north of London, and we packed up some 35 large boxes of papers, and had these shipped to La Jolla.

A few years later, Griffith's daughter, Harriet Tuckey, decided to write an account of her father's role in the first ascent of Everest, and the resulting book became a best seller.[6] Naturally she wanted to find as much of her father's material as she could, and she was taken aback to learn that it was all 5000 miles away. However she visited UCSD and spent several weeks going through the papers and photographing many of them. At the end of that time she told me that initially she was very put out to learn that all the documents were so far away, but after studying them, she was actually very pleased, because she recognized that they would be kept for perpetuity under ideal conditions.

The archive contains material on high altitude medicine and physiology from a large number of sources. One of these is Ulrich Luft who had a particularly interesting history. He was German, and during World War II, he worked in a laboratory in Berlin on high altitude medicine problems. Then after the war, he moved to the United States and eventually took a senior position in the Lovelace Foundation in Albuquerque, NM. I knew his work well, and some time after he died, his son who is an academic nephrologist came to a meeting in San Diego, and I

[5]There is information about the Archival Collection in High Altitude Medicine and Physiology at http://roger.ucsd.edu/search/t?SEARCH=high-altitude+medicine+and+physiology+collection

[6]Everest—The First Ascent: The Untold Story of Griffith Pugh, the Man Who Made It Possible by Harriet Tuckey, Rider, 2013.

asked him where his father's papers were. He said that they were still in Albuquerque in his mother's home but she was confused, and he thought that she would probably throw them away. We contacted her, and the upshot was that the UCSD library sent a van to her home and collected a trove of material. This is a particularly interesting collection because it includes work done in Berlin before and during World War II.

We have many other collections and they can be accessed on-line through the UCSD Library catalog. A very interesting group of papers came from Charles Houston who was a doyen of high altitude medicine some years ago, and gave one of the first descriptions of high altitude pulmonary edema. We also have material from Thomas Hornbein who was a member of the team that made the first American ascent of Everest in 1963. He reached the summit via the West ridge and then descended using the Southeast route, thus traversing the mountain which was a major achievement. He and several other climbers were stopped by darkness during the descent and had to spend the night holed up in the snow not far from the summit. They had no bottled oxygen and were extremely fortunate to survive.

The archival collection continues to grow. For example, we recently received a large collection of papers from Sukhamay Lahiri who was a member of both the Silver Hut and AMREE expeditions. He was well known for his research on oxygen sensing by the peripheral chemoreceptors. One of the nice things about an archival collection is that it continues to become more valuable as time goes on.

11.5 Oxygen Conditioning

To conclude this section, a few words will be added about three ongoing research projects. The first of these interests me greatly at the present time. Oxygen conditioning is like air-conditioning, but whereas that procedure changes the temperature and perhaps the humidity of the air that is being circulated through a building, oxygen conditioning changes its oxygen concentration. This has not yet been done on a large scale, but I believe that it has an enormous potential for improving the well-being and productivity of people at high altitude.

My interest in this topic was sparked by an invitation in 2015 to contribute to a special issue of the (British) Journal of Physiology that was devoted to the physiologist, Joseph Barcroft. I decided to write an article prompted by his bold statement "All dwellers at high altitude are persons of impaired physical and mental powers". By dwellers, Barcroft meant everybody located at high altitude, and by impaired he meant compared with being at sea level.

Now there is extensive evidence that all visitors to high altitude have a reduced exercise tolerance, and also that, if the altitude is high enough, they have impaired neuropsychological function. As far as we know, the same applies to so-called sojourners. These are people from low altitude who find themselves working at high altitude, often for several years. The really interesting group is the permanent residents of high altitude. There are millions of these, especially in the South

American Andes, and on the Tibetan plateau. It has often been assumed that these permanent residents, who have frequently been there for generations, are completely adapted to the hypoxic conditions. However there is increasing evidence that this is not the case.

For example, measurements show that permanent residents can increase their maximal oxygen consumption when they descend to a lower altitude. This means to me that they are not fully adapted to the high altitude. It is hardly surprising that they can do more work at a lower altitude because presumably the PO_2 of the exercising muscles is increased. The fascinating new evidence is that permanent residents of high altitude apparently have reduced neuropsychological function compared with matched groups at a lower altitude. These are difficult measurements to make but the evidence is strong. What this means to me is that millions of people who live permanently at high altitude might be improved if the altitude was lower.

Oxygen conditioning can essentially do this. For every 1% that the oxygen concentration of the air is raised, the equivalent altitude is reduced by about 300 m. Thus oxygen conditioning is very powerful. Facilities that would be suitable for oxygen conditioning include schools, hospitals, corporate buildings where important decisions are being made, embassies, and perhaps department stores. Of course all these places are air-conditioned in hot climates.[7]

Is oxygen conditioning on a large scale economically feasible? This remains to be seen but the engineering procedure is similar in some respects to air-conditioning. In both instances the major process is compressing gas. In oxygen conditioning the compressed air is passed through a synthetic zeolite that preferentially absorbs the nitrogen. In air-conditioning the air is passed through a chiller that works by compressing a refrigerant gas. I believe that if the gains are substantial, technology will solve the problem.

Oxygen conditioning is an extension of a previous procedure that we developed over 20 years ago. This consists of raising the oxygen concentration of individual rooms at high altitude using small oxygen concentrators. This process is known as oxygen enrichment and is presently used in many facilities at high altitude including mines, dormitories, telescope facilities, and some luxury hotels.[8] Oxygen conditioning is simply taking this technology to a higher level.

[7]West, J.B. Oxygen conditioning: a new technique for improving living and working at high altitude. *Physiology (Bethesda)* 31: 216-212, 2016.

[8]West, J.B. Oxygen enrichment of room air to relieve the hypoxia of high altitude. *Resp. Physiol.* 99: 225-232, 1995.

11.6 Assessing Pulmonary Gas Exchange in Patients by Analyzing Expired Gas

A remarkable thing about this project is that I have come full circle. The first article I ever published (in 1957) was on the measurement of ventilation-perfusion inequality from a single expiration.[9] So here we are again. One difference is that 60 years ago the measurements were made with a respiratory mass spectrometer the size of a refrigerator (see Fig. 4.1). Now miniature gas analyzers are available, and the equipment can be carried in one hand.

Of course the main advantage of using expired gas as opposed to arterial blood is that the procedure is completely non-invasive. It is too early to judge how useful this will be in clinical practice. However the end-tidal PCO_2 is probably a useful measure of the arterial value in many patients. If we combine the expired gas analysis with a pulse oximeter, we can get a useful estimate of the arterial PO_2 in many circumstances. It also appears that the end-tidal PO_2 may be a useful measure of the alveolar gas value. If so, this means that a measurement similar to the traditional alveolar-arterial difference may be possible without taking arterial blood. However the project is in its early days.

11.7 Pulmonary Function in Pilots of High-Performance Fighter Aircraft

Pilots of high-performance aircraft occasionally develop short periods of confusion. The reason for this is unknown. Hypoxia is suspected but not yet proved. The problem is serious, and at least one pilot with his aircraft has been lost as a result. I have taken part in long discussions, set up by Boeing, about the problem in relation to the F22 fighter aircraft. These telecoms have included pilots, engineers, and other experts but as yet there has been no clear outcome. The pilots train by taking part in simulated combat maneuvers, during which they are exposed to many stresses including high G levels and rapid changes in cabin altitude, and possibly the problem is related to some combination of these stresses.

Now here is an extraordinary situation. Hundreds of measurements are made on the function of different systems of the aircraft on every flight, but not one on the pilot. The only exception is a pulse oximeter in some situations. This is bizarre. We are therefore developing equipment that will allow simple measurements of pulmonary gas exchange of the pilot during flight. Here it is essential that the test equipment dies not interfere with the pilot in any way. A capillary tube from the

[9] West, J.B., K.T. Fowler, P. Hugh-Jones and T.V. O'Donnell. Measurement of the ventilation-perfusion ratio inequality in the lung by the analysis of a single expirate. *Clin. Sci.* 16: 529-547, 1957.

base of the pilot's mask leads to miniature oxygen and carbon dioxide analyzers in the pilot's vest so that the inspired and expired PO_2 and PCO_2 can be continually measured. In addition the pressure in the mask and the inspired flow rate are monitored.[10] The project, which is being developed with the US Air Force, is in its early stages but shows great promise.

[10]West, J.B. A strategy for in-flight measurements of physiology of pilots of high-performance fighter aircraft. *J Appl Physiol.* 115:145-149, 2013.

Chapter 12
Commentary on Some of My Books and Scientific Papers

Part 1 of this Memoir is a narrative of my career, and details of the science were intentionally omitted so that readers who do not have a scientific background would find the account accessible. However some readers may wish to have further information about the science, and so here is a list of some of my most important books and articles. The published abstracts or summaries have been included where these are available, together with brief comments on the significance of the work. The numbers refer to the complete list of my publications.

Books

1. West, J.B., *Ventilation/Bloodflow and Gas Exchange*, Oxford: Blackwell Scientific Publications; Philadelphia: Lippincott, 1st edition, 1965. I wrote my first book about ventilation-perfusion relationships in the lung because I found the topic so difficult to understand. The book had steady sales for some 30 years, particularly in the U.K., and went through five editions.
3. West, J.B., *Respiratory Physiology – The Essentials*, Baltimore: Williams & Wilkins, 1st edition, 1974. This is my best known book and has been translated into 15 languages. At least two of the translations (into Chinese and Farsi) were pirated (see Fig. 8.4). The book is used by medical students throughout the world, and is the only one that has made any money worth talking about. The present edition is the 10th and Andrew M Luks joined me as co-author.
4. West, J.B., *Translations in Respiratory Physiology*, Stroudsburg: Dowden, Hutchinson and Ross, 1975. This was a very satisfying project. It was partly prompted by a colleague, Lee Langley, who told me there was money available through the National Library of Medicine to support publishing ventures outside the U.S. This was some sort of pay-back for loans. The translations of classical papers in respiratory physiology were done by people in Israel and

The online version of this chapter (doi:10.1007/978-1-4939-7122-0_12) contains supplementary material, which is available to authorized users.

J.B. West, *Breathing on the Roof of the World*, Springer Biographies,
DOI 10.1007/978-1-4939-7122-0_12

the pages were printed there but the book was published in the U.S. Many of the papers were classics from the late nineteenth or early twentieth century and the book remains a valuable resource though copies are hard to come by.

5. West, J.B. (ed.), *Regional Differences in the Lung*, New York: Academic Press, 1977, This book summarizes the work that my group did at Hammersmith on the effects of gravity on the lung with some additional invited contributions.
7. West, J.B., *Pulmonary Pathophysiology – The Essentials*, Baltimore: Williams & Wilkins, 1st edition, 1977. This is a companion to *Respiratory Physiology – The Essentials* and deals with the diseased as opposed to the normal lung. It continues to be used but is not as popular as *Respiratory Physiology* partly because of the way the subject is taught in medical schools. The 9th edition was published in 2017 and has Andrew M Luks as co-author
11. West, J.B. (ed.), *Best & Taylor's Physiological Basis of Medical Practice*, 11th edition, Baltimore, MD: Williams & Wilkins, 1985. I was flattered to be asked to edit this classical physiology textbook because I used an earlier edition when I was a medical student nearly 40 years previously. I edited two editions with most of the contributions coming from faculty at UCSD. Williams & Wilkins then decided to drop the book because they argued that it was too encyclopedic but they later regretted this. Although the 12th edition came out in 1991and is therefore way out of date I still occasionally get emails from medical students about it.
12. West, J.B., *Everest – The Testing Place*, New York: McGraw-Hill, 1985. This is the story of the 1981 American Medical Research Expedition to Everest and I still get a buzz from reading it, particularly the debriefings of the three Americans who summited, and the transcripts of their radio links. Unfortunately McGraw-Hill remaindered it after a year or two without telling me. I could have bought a number of copies for a dollar each. Copies can still be found on the Internet.
14. J.B. West, R.B, Schoene, A. M. Luks and J.S. Milledge, *High Altitude Medicine and Physiology*, London: CRC Press, edition, 5th edition. This began as the book Mountain Medicine by Michael Ward, published in 1975. Jim Milledge and I then joined him as co-authors for later editions, and it has become a standard medical textbook for physicians who want information about high-altitude medicine. It is now in its 5th edition.
15. Crystal R.G., J.B. West (chief eds.), P.J. Barnes, N.J. Cherniack, E.R. Weibel (assoc. eds.), *The Lung: Scientific Foundations*, (2 vols.). New York: Raven Press, 1st edition, 1991. This was a very influential book when it first appeared, and the 2nd edition published in 1997 remains a useful comprehensive book on the basic science of respiratory medicine.
17. West, J.B. (ed.), *Respiratory Physiology: People and Ideas*, New York: Oxford University Press, 1996. This book is composed of chapters written by eminent physiologist who worked on respiratory physiology in the thirty or so years after World War II. It had very good reviews.
19. West, J.B., *High Life: A History of High-Altitude Physiology and Medicine*, New York: Oxford University Press, 1998. This is my most scholarly book

and aims to be a definitive history of high altitude medicine and physiology. It is cited by booksellers dealing in old books who want to refer to the historical setting of books in their catalogues.

21. Prisk, G.K., M. Paiva, and J.B. West (eds.), *Gravity and the Lung: Lessons from Microgravity*, New York: Dekker, 2001. This book summarizes our work in the NASA program on the changes of lung function in space. It also includes contributions from other groups
22. West, J.B. *Essays on the History of Respiratory Physiology*. Washington D.C. American Physiological Society/Springer 2015. This is a collection of some of my historical articles that have been previously published in various journals.

Scientific Articles

1. West, J.B., K.T. Fowler, P. Hugh-Jones and T.V. O'Donnell, Measurement of the ventilation-perfusion ratio inequality in the lung by the analysis of a single expirate. *Clin. Sci.* 16: 529–547, 1957.

Comment: This was my first publication. It described experiments in which we analyzed expired gas from human volunteers using the newly built mass spectrometer, and so obtained an index of ventilation-perfusion inequality in the lung. Remarkably we used the same experimental procedure to study astronauts in the International Space Station some 40 years later.

4. Dyson, N.A., P. Hugh-Jones, G.R. Newbery and J.B. West, The preparation and use of oxygen-15 with particular reference to its value in the study of pulmonary malfunction. *Second United Nations International Conference on the Peaceful Uses of Atomic Energy*, Geneva, 1958.

Comment: This was the first report of our work with radioactive oxygen-15. This gas has a half-life of only two minutes and so it had to be continuously produced using the newly completed cyclotron on the hospital campus while we were performing the experiments.

14. Hugh-Jones, P. and J.B. West, Detection of bronchial and arterial obstruction by continuous gas analysis from individual lobes and segments of the lung. *Thorax* 15: 154–164, 1960.

Summary: The technique of continuous gas analysis from individual lobes and segments of the lungs during diagnostic bronchoscopy is described. This is achieved by inserting a sampling tube of a special mass spectrometer into the regional bronchus. The gas analysis is presented on a four-channel direct-writing recorder. It shows how simultaneous records of the local changes in the oxygen, carbon dioxide, and argon (following introduction of argon into the bronchial tree) give characteristic patterns from which the presence of partial bronchial obstruction, arterial obstruction, or a combination of both can be inferred.

Comment: This paper described the measurement of regional lung function during bronchoscopy by passing the sampling tube of the mass spectrometer into various parts of the lung. The clinical measurements were supplemented by studies on anesthetized dogs. This was a novel, enterprising project but it did not lead to any important advances.

15. West, J.B. and C.T. Dollery, Distribution of blood flow and ventilation-perfusion ratio in the lung, measured with radioactive CO_2. *J. Appl. Physiol.* 15: 405–410, 1960.

Abstract: Inhaled radioactive CO_2 is rapidly taken up by pulmonary blood. By external counting over the chest during breath holding, the clearance rate of radioactive CO_2 from the counting field can be recorded, and is proportional to the regional perfusion. In normal subjects, the clearance rate varied from about 20%/sec. at the base of the lung to virtually nil at the apex, and the change was approximately linear with distance up the chest. The difference between upper and lower zones was reduced on moderate exercise and eliminated when the subject lay on his back. By relating the counting rate at the end of inspiration to the volume of lung in the counting field, the difference in ventilation between upper and lower zones was measured and found to be small. Variation in ventilation-perfusion ratio was thus determined. Alveolar-arterial O_2 gradient expected from this ventilation-perfusion ratio inequality was calculated to be about 4 mm Hg. This suggests that the variation in blood flow between upper and lower parts of the lung in erect man accounts for the whole of the ventilation-perfusion ratio inequality in the normal lung.

Comment: This was the first clear demonstration of the striking regional differences of blood flow in the human lung caused by gravity. The study received a lot of attention when it was published and the paper is still frequently cited.

16. West, J.B., Measurement of bronchial air flow. *J. Appl. Physiol.* 15: 976–978, 1960.

Comment: This described a method for measuring the rate of air flow in single airway of the lung. The device consisted of two long fine tubes (hypodermic steel tubing) with right-angled ends. These were apposed with a small separating gap. A tracer gas, for example argon, was blown down one tube, and the other tube was connected to a vacuum pump and gas analyzer. The argon streaming across the gap was deflected by the air flow at right angles to it, and the argon signal from the analyzer was a measure of the air flow rate. In practice the device was sited in a small bronchus during bronchoscopy, and the analyzer was a mass spectrometer. One of the most interesting findings was the amount of the pulsatility of flow caused by movement of the heart. I subsequently patented the invention but heard no more about it.

23. West, J.B. and C.T. Dollery. Absorption of inhaled radioactive water vapour. Nature 189: 588, 1961.

Comment: This brief Letter to Nature reported a surprising finding. We prepared a bag of water vapor labeled with oxygen 15 and took a deep breath from the bag. To our astonishment, radiation counters situated over the lungs showed no response. However when we moved the counters to the region of the pharynx, there was a high counting rate. This showed how effectively the water in the inspired gas was exchanged with the water lining the upper respiratory tract. The finding is consistent with the rapid saturation of inspired gas with water vapor at body temperature in the upper respiratory tract.

30. West, J.B., Diffusing capacity of the lung for carbon monoxide at high altitude. *J. Appl. Physiol.* 17: 421–426, 1962.

Abstract: Diffusing capacity of the lung for carbon monoxide (Dl) was measured in London (Pb approx. 750 mm Hg), at 15,300 ft (4700 m; Pb approx. 440 mm Hg), and at 19,000 ft (5800 m; Pb approx. 380 mm Hg) on seven members of the Himalayan Scientific and Mountaineering Expedition, 1960–61. At each altitude Dl was measured at two work levels (300 and 900 kg-m/min) and at three different inspired oxygen tensions in order to separate membrane and blood components of the diffusing barrier. A steady state method was used with mixed expired gas analysis; dead space-to-tidal volume ratio was assumed but calculated Dl was insensitive to this. There was no consistent change in Dl at 15,300 f. (subjects breathing ambient air) compared with sea level, but Dl was significantly increased after 7–10 weeks at 19,000 f. (mean changes of 15 and 19% for work levels of 300 and 900 kg-m/min, respectively). However, this small change in Dl can be wholly accounted for by the increased rate of reaction of carbon monoxide with hemoglobin due to hypoxia and by the increased blood hemoglobin concentration.

Comment: This and the subsequent paper, number 31, were two of the most important papers that came out of the Silver Hut Expedition of 1960–1961. The present study was done because of the evidence that the diffusing capacity is likely to be an important limiting factor for exercise at very high altitudes.

31. West, J.B., S. Lahiri, M.B. Gill, J.S. Milledge, L.G.C.E. Pugh and M.P. Ward. Arterial oxygen saturation during exercise at high altitude. J. Appl. Physiol. 17: 617–621, 1962.

Abstract: Arterial oxygen saturations were measured on six members of the Himalayan Scientific and Mountaineering Expedition, 1960–61, during a wintering period at 19,000 f. (5800 m; Pb 380 mm Hg). The determinations were made by ear oximetry and by the analysis of venous blood from the heated hand during rest and exercise at work levels up to 1200 kg-m/min. Expired gas volumes and gas concentrations were also measured. The average arterial oxygen saturation at rest was 67%, and at work levels of 300 and 900 kg-m/min it was 63 and 56%, respectively. Several readings of less than 50% saturation were recorded during severe exercise. The progressive fall in arterial oxygen saturation as the work level was raised occurred in spite of an increasing alveolar oxygen tension, and the resulting large alveolar-arterial oxygen differences can be explained by the diffusion limitations of the lung.

Comment: This article shows the striking fall in arterial oxygen saturation during exercise at very high altitude which can be explained by diffusion limitation across the blood-gas barrier. It is interesting that this is the only situation where diffusion limitation of oxygen transfer across the capillary wall consistently occurs in the normal lung.

40. West, J.B., C.T. Dollery and A. Naimark, Distribution of blood flow in isolated lung; relation to vascular and alveolar pressures. *J. Appl. Physiol.* 19: 713–724, 1964.

Abstract: The left lung from a dog was removed, ventilated with negative pressure, and perfused with venous blood. Pulmonary arterial, venous, and alveolar pressures could be varied over a large range. The distribution of blood flow in the lung was measured with Xe133. Under these conditions, there was no blood flow above the level at which alveolar equaled arterial pressure (measured at the arterial cannula). Below this level there was a linear increase in blood flow down the lung when the venous pressure was kept low. Raising the venous pressure made the distribution of flow more uniform below the level at which venous and alveolar pressures were equal although flow still increased down this zone. The flow distribution could be completely accounted for by the mechanical effects of the pressure inside and outside the blood vessels which each behaved like a Starling resistance. It was possible to simulate the flow distributions found in man in various physiological and diseased states.
Comment: This often cited paper reports the first comprehensive description of the factors responsible for the uneven distribution of blood flow in the lung. The diagram of the three lung zones is reproduced in many textbooks of physiology.

66. Glazier, J.B., J.M.B. Hughes, J.E. Maloney and J.B. West, Vertical gradient of alveolar size in lungs of dogs frozen intact. *J. Appl. Physiol.* 23: 694–705, 1967.

Abstract: Dog lungs were fixed in situ buy freezing and alveolar size was measured by histologic morphometric techniques. In the erect lung at functional residual capacity the apical alveoli were about four times larger by volume than the basal ones, most of the change in size being over the upper 10 cm of lung. The difference between apex and based increased to 11:1 when the animals were exposed to 3G on a centrifuge. No difference in size was found when the lungs were expanded by 30 cm H_2O pressure. In horizontal lungs, alveolar size was the same at the apex and base, but the most superior alveoli were larger than the dependent ones. In inverted dogs alveolar size was uniform from apex to base. The differences in alveolar volume can be explained if the transpulmonary pressure at any level is determined by the cross-sectional area of the lung and the weight of the lung below that level. This pressure apparently changes more rapidly in the upper part of the erect lung than in the lower part.
Comment: This was the first direct demonstration of the uneven expansion of the lung caused by gravity.

77. Glazier, J.B., J.M.B. Hughes, J.E. Maloney and J.B. West, Effect of acceleration on alveolar size in the lungs of dogs. *Aerospace Med.* 39: 282–288, 1968.

Abstract: Alveolar size has been measured in different regions of the lungs of dogs subjected to 3G headwards acceleration (+3Gz) and 5G supine acceleration (+5Gx). The animals were frozen intact on a human centrifuge thereby fixing the lungs while they were exposed to the increased G forces. The measurements were made using histological morphometric techniques. The normal gradient of alveolar size of 3.7 to 1 from apex to base in the vertical dog at 1G increased to 11 to 1 at 3G if the dogs wore an abdominal binder as an antigravity suit. There was no significant change in the gradient of size when the abominable binder was not worn. When exposed to an acceleration of 5G, the most superior alveoli in the supine dog more than doubled their volume. Most of the alveoli more than 9 cm below the ventral surface of the dog were collapsed by the G forces

Comment: This reports a remarkable experiment, which was a tour de force, where anesthetized dogs were accelerated on a human centrifuge for several hours while their lungs were frozen by surrounding the animal with frozen CO_2. The results are relevant to the distortion of the lung that presumably occurs in pilots of high performance aircraft during high G turns. A bizarre feature of the publication was that only the first author was originally listed because of an editing error.

89. Glazier, J.B., J.M.B. Hughes, J.E. Maloney and J.B. West, Measurements of capillary dimensions and blood volume in rapidly frozen lungs. *J. Appl. Physiol.* 26: 65–76, 1969.

Abstract: Isolated perfused dog lungs were rapidly frozen with liquid Freon. Measurements of the number of red blood cells per 10 μ of alveolar septum, the mean width of capillaries, and the percent of septum occupied by red blood cells were made in freeze-dried sections. The effects of changing vascular and alveolar pressures were studied, with particular attention to the relationship of downstream pressure to alveolar pressure; that is venous pressure was set lower than alveolar pressure (zone II) or higher (zone III). When alveolar pressure exceeded arterial pressure there were less than 0.5 red blood cells per 10 μ of septum and mean capillary width was 2.3 μ. As capillary pressure increased the number of red blood cells per 10 μ of septum increased to a maximum of 2.8 at a capillary pressure of 50 cm H_2O. Mean capillary width increased to 6.5 μ. No further changes occurred in either measurement at pressures up to 100 cm H_2O. At high lung volumes both red blood cell density and capillary width were greatly reduced. Recruitment of new vessels was the predominant mechanism to account for the increase in red blood cell concentration down zone II, whereas expansion of already open capillaries was more important in zone III.

Comment: In this paper, pulmonary capillaries near the surface of the lung were rapidly frozen to clarify the relations between structure and function, particularly the effects of distension of the capillaries. The results have many implications for understanding the physiology of pulmonary capillaries.

93. West, J.B., Ventilation-perfusion inequality and overall gas exchange in computer models of the lung. *Resp. Physiol.* 7: 88–110, 1969.

Abstract: The effects of increasing ventilation-perfusion inequality on overall gas exchange were studied in digital computer models of the lung. Ventilation/unit volume and perfusion/unit volume were distributed log normally with respect to lung volume. Ventilation-perfusion inequality was found to affect the transfer of carbon dioxide nearly as much as oxygen. The reasons for the misconception that inequality does not interfere with carbon dioxide transfer are discussed. The effects of changing overall ventilation, blood flow, inspired oxygen, hemoglobin concentration, and the acid-based status of the blood were investigated when oxygen uptake and carbon dioxide output were held constant at normal values. In general, mismatch of ventilation and blood flow in a lung cause the arterial oxygen tension to fall and the carbon dioxide to rise; increasing overall ventilation rapidly restored blood carbon dioxide tensions to normal but improved oxygen tensions little. Considerable improvement in oxygen tensions occurred, however, if the oxygen dissociation curve was made linear. In spite of large increases in inspired oxygen, severe hypoxemia and large alveolar-arterial oxygen differences remained when the inequality was severe. The results help to explain how the lung maintains oxygen and carbon dioxide transfer in the face of ventilation-confusion ratio inequality and may be useful in characterizing the degree of ventilation-perfusion inequality in real situations.

Comment: This novel paper showed how modern computing techniques can elucidate pulmonary gas exchange. It was helpful in the later development of the MIGET technique.

103. West, J.B., Causes of carbon dioxide retention in lung disease. *New Engl. J. Med.* 284: 1232–1236, 1971.

Comment: This article in the influential *New England Journal of Medicine* dealt with an important misconception about the causes of carbon dioxide retention in chronic lung disease.

106. Denison, D.M., D.A. Warrell and J.B. West, Airway structure and alveolar emptying in the lungs of sea lions and dogs. *Resp. Physiol.* 13: 253–260, 1971.

Abstract: The pressure-volume characteristics of the lungs excised from four California sea lions (*Zalophus californianus*) and six dogs have been compared by spirometry during cyclic changes of inflation pressure from + 30 to−30 cm H_2O and histologically after recompression to atmospheric pressure from an absolute pressure of 112 mm Hg. In comparison with the dog lungs, the sea lion lungs emptied more completely on mild compression and much more completely on severe compression. These findings support Scholander's hypothesis that some marine mammals are protected from decompression sickness by cartilaginous reinforcement of the small airways which permits alveolar emptying during a dive, so isolating compressed gas from pulmonary capillary blood.

Comment: An entertaining study showing that the lungs of sea lions can empty more completely than those of dogs. This behavior is important in protecting sea lions from decompression sickness.

108. West, J.B. and F.L. Matthews, Stresses, strains, and surface pressures in the lung caused by its weight. *J. Appl. Physiol.* 32: 332–345, 1972.

Abstract: In an effort to understand how the lung is deformed by its own weight, we have analyzed the distribution of regional expansion, stresses, and the surface pressures in a theoretical elastic lung-shaped model using the technique of finite elements. In the upright posture, the parenchyma was most expanded at the apex and least at the base. Stresses in both the vertical and lateral direction were maximal at the apex. As the lung was inflated from very low volumes to total lung capacity, parenchymal expansion and stress at the apex first decreased then increased. This behavior can be explained by the increasing rigidity of the expanded lung which enabled it to resist distortion by its own weight. At functional residual capacity, the stress at the apex was near its minimum. The differences in intrapleural pressure down the lung were volume dependent, increasing at very low volumes. In the inverted lung, the regional distributions in stress, strain, and surface pressures were less marked because of the shape of the chest.

Comment: This was an innovative theoretical study using a sophisticated engineering technique known as finite element analysis and was done in conjunction with the Department of Aeronautics at Imperial College in London.

116. Evans, J.W., P.D. Wagner and J.B. West, Conditions for reduction of pulmonary gas transfer by ventilation-perfusion inequality. *J. Appl. Physiol.* 36: 533–537, 1974.

Abstract: Although it has been claimed that CO_2 elimination is not impaired by ventilation-perfusion inequality because of its almost linear dissociation curve, CO_2 retention has been observed in lung models. Here we show formally that mismatching of ventilation and blood flow must reduce the pulmonary elimination or uptake of any gas with a linear dissociation curve, irrespective of the pattern of ventilation-perfusion ratios. We define the properties which a gas must have for it to be possible for ventilation-perfusion inequality to enhance pulmonary elimination or uptake, and a theoretical example is given. This condition is not met for any known gas under physiological steady-state conditions

Comment: I think this paper is a gem but it is essentially unknown. It owed much to the mathematical expertise of John Evans who is an M.D. but a Professor of Mathematics at UCSD. It describes a theoretical situation where uneven ventilation and blood flow can actually improve gas exchange.

117. Wagner, P.D., H.A. Saltzman and J.B. West, Measurement of continuous distributions of ventilation-perfusion ratios: theory. *J. Appl. Physiol.* 36: 588- 599, 1974.

Abstract: Most previous descriptions of the distribution of ventilation-perfusion ratios (VA/Q) divide the lungs into only two or three uniform compartments. However, an analysis which would result in definition of the position, shape, and dispersion of the distribution would be more realistic. We describe here such a technique, applicable both in health and disease, in which the characteristics of distributions containing up to three modes can be determined. In particular, areas with low but finite VA/Q ratios are separated from areas whose VA/Q is zero (shunt), and regions with high VA/Q ratios are differentiated from regions that are nonperfused (dead space). To perform the measurement, dextrose solution or saline is equilibrated with a mixture of several gases of different solubilities, and then infused into a vein. After a steady state has been established, the concentration of each gas are measured in the mixed arterial blood and mixed expired gas. The curve relating arterial concentration and solubility is transformed into a virtually continuous distribution of blood flow against VA/Q, using techniques of numerical analysis. The relation between expired concentration and solubility is similarly converted into the distribution of ventilation. The numerical analysis technique has been tested against many artificial distributions of VA/Q ratios and these have all been accurately recovered.

Comment: This is the first paper describing the theoretical basis of the multiple inert gas elimination technique (MIGET) developed by my colleague, Peter Wagner. This was a breakthrough in understanding the mechanism of impaired gas exchange in patients with lung disease.

122. Wagner, P.D., R.B. Laravuso, R.R. Uhl and J.B. West, Continuous distributions of ventilation-perfusion ratios in normal subjects breathing air and 100% O_2. *J. Clin. Invest.* 54: 54–68, 1974.

Abstract: A new method has been developed for measuring virtually continuous distributions of ventilation-perfusion ratios (V(A)/Q) based on the steady state elimination of six gases of different solubilities. The method is applied here to 12 normal subjects, aged 21–60. In nine, the distributions were compared breathing air and 100% oxygen, while in the remaining three, effects of changes in posture were examined. In four young semirecumbent subjects (ages 21–24) the distributions of blood flow and ventilation with respect to V(A)/Q were virtually log-normal with little dispersion (mean log standard deviations 0.43 and 0.35, respectively). The 95.5% range of both blood flow and ventilation was from V(A)/Q ratios of 0.3–2.1, and there was no intrapulmonary shunt (V(A)/Q of 0). On breathing oxygen, a shunt developed in three of these subjects, the mean value being 0.5% of the cardiac output. The five older subjects (ages 39–60) had broader distributions (mean log standard deviations, 0.76 and 0.44) containing areas with V(A)/Q ratios in the range 0.01–0.1 in three subjects. As for the young subjects, there was no shunt breathing air, but all five developed a shunt breathing oxygen (mean value 3.2%), and in one the value was 10.7%. Postural changes were generally those expected from the known effects of gravity, with more ventilation to high V(A)/Q areas when the subjects were erect than supine. Measurements of the shunt while breathing oxygen, the Bohr CO(2) dead space, and the alveolar-arterial oxygen difference were all consistent with the observed distributions. Since the method involves only a short infusion of dissolved inert gases, sampling of arterial blood and expired gas, and measurement of cardiac output and minute ventilation, we conclude that it is well suited to the investigation of pulmonary gas exchange in man.

Comment: The first paper describing the application of the multiple inert gas elimination technique (MIGET) to normal volunteers

135. Dantzker, D.R., P.D. Wagner and J.B. West, Instability of lung units with low VA/Q ratios during O2 breathing. *J. Appl. Physiol.* 38: 886–895, 1975.

Abstract: Using a multiple inert gas infusion method, we have observed the development of shunts during oxygen breathing in lungs which contained areas of low ventilation-perfusion ratios while breathing air. This paper gives a theoretical analysis of the factors involved. When the inspired ventilation-perfusion ratio (VaI/Q) of a lung unit is gradually reduced, a point is reached where the expired ventilation falls to zero. Such a unit will no longer eliminate gas but may continue gas uptake unless it becomes atelectatic. This critical VaI/Q is determined by the net flux of O_2, CO_2, and N_2 from alveolar gas to capillary blood, and its value increases from about 0.001 to 0.1 as the inspired gas is changed from air to 100% O_2. The critical VaI/Q at any inspired O_2 concentration is raised if the O_2 or N_2 content of mixed venous blood are reduced or if N_2 is replaced by a more soluble gas. In distributions of ventilation-perfusion ratios, the amount of shunt which develops during oxygen breathing depends on the degree of dispersion of the VaI ratios. The release of hypoxic vasoconstriction following O_2 administration, in general, reduces the amount of shunt.

Comment: An important study showing the mechanism by which oxygen breathing can cause shunts in the lung.

144. Wagner, P.D., D.R. Dantzker, R. Dueck, J.L. Clausen and J.B. West. Ventilation-perfusion inequality in chronic obstructive pulmonary disease. J. Clin. Invest. 59: 203–216, 1977.

Abstract: A multiple inert gas elimination method was used to study the mechanism of impaired gas exchange in 23 patients with advanced chronic obstructive pulmonary disease (COPD). Three patterns of ventilation-perfusion (VA/Q) inequality were found: (a) A pattern with considerable regions of high (greater than 3) VA/Q, none of low (less than 0.1) VA/Q, and essentially no shunt. Almost all patients with type A COPD showed this pattern, and it was also seen in some patients with type B. (b) A pattern with large amounts of low but almost none of high VA/Q, and essentially no shunt. This pattern was found in 4 of 12 type B patients and 1 of type A. (c) A pattern with both low and high VA/Q areas was found in the remaining 6 patients. Distributions with high VA/Q areas occurred mostly in patients with greatly increased compliance and may represent loss of blood-glow due to alveolar wall destruction. Similarly, well-defined modes of low VA/Q areas were seen mostly in patients with severe cough and sputum and may be due to reduced ventilation secondary to mechanical airways obstruction or distortion. There was little change in the VA/Q distributions on exercise or on breathing 100% O_2. The observed patterns of VA/Q inequality and shunt accounted for all of the hypoxemia at rest and during exercise. There was therefore no evidence for hypoxemia caused by diffusion impairment. Patients with similar arterial blood gases often had dissimilar VA/Q patterns.

As a consequence the pattern of VA/Q inequality could not necessarily be inferred from the arterial PO_2 and PCO_2.

Comment: The first article describing the use of MIGET to measure the distribution of ventilation-perfusion ratios in COPD. It had a major impact on the field. The different patterns in Type A and Type B COPD have been very useful in teaching about ventilation-perfusion inequality.

147. West, J.B., State of the art: ventilation-perfusion relationships. *Am. Rev. Resp. Dis.* 116: 919–943, 1977.

Comment: This was a comprehensive review of the physiology of ventilation-perfusion inequality at the time, and is still useful for people who want an introduction to this difficult topic.

156. Michels, D.B. and J.B. West, Distribution of pulmonary ventilation and perfusion during short periods of weightlessness. *J. Appl. Physiol.* 45: 987–998, 1978.

Abstract: Information on the distributions of pulmonary ventilation and perfusion was obtained from four subjects on board a Learjet during 112 weightless periods lasting up to 27 s each. Zero gravity (G) was obtained during all or part of each test by varying the aircraft flight profile. Single-breath N_2 washouts were performed with the test inspiration containing an initial bolus of argon at residual volume (RV). When the test inspiration was at 0 G, and the washout at 0 G or greater, the terminal rises and the cardiogenic oscillations in both N_2 and argon were small and often absent. If instead the test inspiration was at 1 G with the washout at 0 G, the terminal rises were again small or absent but the cardiogenic oscillations remained. The terminal rise and the cardiogenic oscillations for N_2, but not argon, were also nearly eliminated by performing just the preliminary exhalation to RV at 0 G with the test inspiration and washout following at 1 G. Aleveolar plateaus for N_2 sloped upward at 0 G apparently due to nontopographical inequalities of ventilation. In further tests during air breathing, recordings were made of expired partial pressure of oxygen PO_2) and carbon dioxide (POO_2) following a brief hyperventilation and a 15-s breath hold. These recordings revealed marked cardiogenic oscillations in PO_2 and PCO_2 at 1 G that were enhanced at 2 G but almost eliminated at 0 G. The results suggest that virtually all the topographical inequality of ventilation, blood flow, and lung volume seen under 1-G conditions are abolished during short periods of 0 G

Comment: These were our first measurements of pulmonary function during weightlessness and were made in a Learjet aircraft flying a parabolic profile. The changes in weightlessness were dramatic and the paper made a great impression.

163. Mazzone, R., C. Durand and J.B. West, Electron microscopic appearances of lung rapidly frozen under controlled physiological conditions. *J. Applied Physiology* 45:325–333, 1978.

Abstract: Light microscopy of lung rapidly frozen under controlled physiological conditions has been very successful in correlating pulmonary structure and function. However, to study some aspects of pulmonary capillary morphology, the higher resolution of electron microscopy (EM) is necessary. To date, most EM of lung has involved the instillation of a fixative through the airways or vascular system, techniques that probably alter the normal pressure relationships of the capillaries and therefore their morphology. We describe here a technique for rapidly freezing lung to a depth of 1–2 mm below the pleural surface and preparing sections for EM. Lungs from open-chest rats were frozen at various transpulmonary pressures with cold (−80 °C) 70% ethylene glycol. Small pieces were then fixed with a solution containing glutaraldehyde and paraformaldehyde for 24 h at −50 °C. Staining was with osmium tetroxide and uranyl acetate. Lung frozen at high volumes showed marked stretching of the alveolar septa with severe deformation of the capillaries. Lung frozen at low inflation pressures revealed open capillaries containing numerous red blood cells; in addition, infolding of the alveolar wall was frequently seen. We conclude that this technique gives a level of preservation of rapidly frozen lung suitable for electron microscopy

Comment: This demonstration of electron microscopy using rapidly frozen lung was far ahead of its time but sadly was attacked so savagely by the establishment that the first author subsequently abandoned medical research.

164. Michels, D.B., P.J. Friedman and J.B. West, Radiographic comparison of human lung shape during normal gravity and weightlessness. *J. Appl. Physiol.* 47: 851–857, 1979.

Abstract: Human lung shape was measured during zero gravity (0 G) to decide whether the normal vertical regional differences in ventilation are due directly to distortion of the elastic lung by its own weight, or instead, due indirectly to the effect of gravity on the shape of the rib cage and diaphragm. This was important because we previously established that weightlessness virtually abolishes the normal topographical inequality of ventilation (J. Appl. Physiol.: Respirat. Environ. Exercise Physiol. 45: 987–998, 1978). Chest radiographs were made after 10 s of a weightless flight trajectory aboard a NASA-Ames Research Center Learjet in both posterior-anterior and left lateral projections on five seated volunteers at residual volume, functional residual capacity, and total lung capacity. Lung shape was assessed by measuring lung heights and widths in upper, middle, and lower lung regions. We found no significant differences between any of the normal gravity (1 G) and o G measurements, although there was a slight tendency for the lung to become shorter and wider at o G (mean changes generally less than 3% or about 0.5 cm). By contrast, Grassino et al. (J. Appl. Physiol. 39: 997–1003, 1975) found no change in the vertical distribution of ventilation after voluntarily changing lung dimensions by more than 1 cm by moving the abdomen in or out. We conclude that gravity produces the topographical distribution of ventilation in the upright human lung by distorting the elastic lung tissue within the chest rather than by altering the shape of the rib cage and diaphragm.

Comment: This article reported the first chest radiographs taken during weightlessness. There was little change in the shape of the ribcage and diaphragm confirming that the uneven ventilation caused by gravity results from distortion of the lung by its weight.

168. West, J.B. and P.D. Wagner, Predicted gas exchange on the summit of Mt. Everest. *Resp. Physiol.* 42: 1–16, 1980.

Abstract: The ascent of Mt. Everest (altitude 8848 m) by two climbers in May 1978 without supplementary oxygen has prompted us to make a theoretical analysis of gas exchange under these conditions of extreme hypoxia. On the basis of previous measurements made up to an altitude of 7440 m and other data, we have calculated a barometric pressure on the summit of 250 Torr, alveolar PCO_2 of 10 Torr, and Hb concentration of 20.5 g/100 ml. Values for cardiac output, pulmonary capillary blood volume, and diffusing capacity were based on measurements made at 5800 m. A striking result of calculations of oxygenation along the pulmonary capillary is that, even at rest, there is an alveolar-end capillary PO_2 of about 6 Torr caused by diffusion limitation, and this widens rapidly on mild exercise. Arterial and mixed venous PO_2 fall precipitously as the VO2 is raised further. If we assume that the PO2 in mixed venous blood cannot fall below 15 Torr, a maximal VO2 of less than 700 ml/min is predicted. Arterial 02 saturation and, to a smaller extent, the mixed venous PO2 can be increased by shifting the 02 dissociation curve to the left. This can be accomplished by fully compensating for the respiratory alkalosis at a lower altitude, and then climbing rapidly to the summit. Maximal VO2 is extremely sensitive to barometric pressure, and to a lesser extent to lung diffusing capacity. The results are in general agreement with extrapolations from measurements of maximal VO2 at altitudes up to 7440 m.

Comment: This modeling study had an important influence on the design of the 1981 AMREE expedition. For example it emphasized the potential importance of measuring the PO2 and PCO2 in alveolar gas, and also the maximum VO2 for the very low inspired PO2 on the summit.

194. West, J.B., P.H. Hackett, K.H. Maret, J.S. Milledge, R.M. Peters, Jr., C.J. Pizzo and R.M. Winslow, Pulmonary gas exchange on the summit of Mt. Everest. *J. Appl. Physiol.: Resp. Environ. Exercise Physiol.* 55: 678–687, 1983.

Abstract: Pulmonary gas exchange was studied on members of the American Medical Research Expedition to Everest at altitudes of 8050 m (barometric pressure 284 Torr), 8400 m (267 Torr) and 8848 m (summit of Mt. Everest, 253 Torr). Thirty-four valid alveolar gas samples were taken using a special automatic sampler including four samples on the summit. Venous blood was collected from two subjects at an altitude of 8050 m on the morning after their successful summit climb. Alveolar CO2 partial pressure (PCO2) fell approximately linearly with decreasing barometric pressure to a value of 7.5 Torr on the summit. For a respiratory exchange ratio of 0.85, this gave an alveolar O2 partial pressure (PO2) of 35 Torr. In two subjects who reached the summit, the mean base excess at 8050 m was −7.2 meq/l, and assuming the same value on the previous day, the arterial pH on the summit was over 7.7. Arterial PO2 was calculated from changes along the pulmonary capillary to be 28 Torr. In spite of the severe arterial hypoxemia, high pH, and extremely low PCO2, subjects on the summit were able to perform simple tasks. The results allow us to construct for the first time an integrated picture of human gas exchange at the highest point on earth.

Comment: This is one of several papers from the AMREE expedition and perhaps the most important. I love the title!

202. West, J.B., Human physiology at extreme altitudes on Mount Everest. *Science* 223: 784–788, 1984.

Abstract: Extreme altitude presents an enormous physiological challenge to the human body because of severe oxygen deprivation. The American Medical Research Expedition to Everest was specifically designed to study man under these conditions, and successfully obtained physiological data above 8000 m, including a few measurements on the summit itself. The results show that man can tolerate the extreme hypoxia only by an enormous increase in ventilation, which results in an alveolar partial pressure of carbon dioxide of 7.5 torr on the summit and an arterial pH of over 7.7. Even so, the arterial partial pressure of oxygen is apparently less than 30 torr, and maximum oxygen uptake is about 1 L per minute. Additional measurements of ventilation, blood physiology, and metabolic and psychometric changes clarified how man responds to this hostile environment.

Comment: This is a comprehensive report of the findings from AMREE which was published in the prestigious journal *Science*. The cover of this issue of the journal had a very nice photograph from the expedition (see Fig. 9.14).

225. West, J.B., Lactate during exercise at extreme altitude. *Fed. Proc.* 45: 2953–2957, 1986.

Abstract: Maximal exercise at extreme altitude results in profound arterial hypoxemia and, presumably, extreme tissue hypoxia. The best evidence available indicates that the resting arterial PO2 on the summit of Mount Everest is about 28 torr and that it falls even further during exercise. Nevertheless, some ten climbers have now reached the summit without supplementary oxygen. Paradoxically, blood lactate for a given work rate at high altitude in acclimatized subjects is essentially the same as at sea level. Because work capacity decreases markedly with increasing altitude, maximal blood lactate also falls. Extrapolation of available data up to 6300 m indicates that a climber who reaches the Everest summit will have no increase in blood lactate. The cause of the low blood lactate during exercise at extreme altitude is not fully understood. One possibility is depletion of plasma bicarbonate in acclimatized subjects, which reduces buffering and results in large increases in H+ concentration for a given release of lactate. The consequent local fall in pH may inhibit enzymes, e.g., phosphofructokinase (EC 2.7.1.56), in the glycolytic pathway.

Comment: This was probably the first publication to refer to the "lactate paradox," that is the unexpectedly low level of blood lactate in acclimatized subjects at high altitude during maximal work. This has not yet been satisfactorily explained.

227. West, J.B. Highest inhabitants in the world. *Nature* 324: 517, 1986.

Comment: This Letter to Nature followed a visit to the highest mine in the world where I met the Bolivian caretakers who were living at an altitude of 5950 m. This was a sulfur mine in north Chile that has since been closed.

230. West, J.B., Alexander M. Kellas and the physiological challenge of Mount Everest. *J. Appl. Physiol.* 63, 3–11, 1987.

Abstract: Alexander M. Kellas (1868–1921) was a British physiologist who made pioneering contributions to the exploration of Everest and to the early physiology of extreme altitudes, but his physiological contributions have been almost completely overlooked. Although he had a full-time faculty position at the Middlesex Hospital Medical School in London, he was able to make eight expeditions to the Himalayas in the first two decades of the century, and by 1919 when the first official expedition to Everest was being planned, he probably knew more about the approaches than anybody else. But his most interesting contributions were made in an unpublished manuscript written in 1920 and entitled "A consideration of the possibility of ascending Mount Everest." In this he discussed the physiology of acclimatization and most of the important variables including the summit altitude and barometric pressure, and the alveolar PO2, arterial oxygen saturation, maximal oxygen consumption, and maximal ascent rate near the summit. On the basis of this extensive analysis, he concluded that "Mount Everest could be ascended by a man of excellent physical and mental constitution in first-rate training, without adventitious aids [supplementary oxygen] if the physical difficulties of the mountain are not too great." Kellas was one of the first physiologists to study extreme altitude, and he deserves to be better known.

Comment: This is one of my first historical articles and was the result of extensive research in the archives of the Royal Geographical Society in London. At the time, almost nothing had been written about Kellas who has a unique place in the history of high altitude physiology. Now his importance is appreciated, and there is also a fine book.

261. West, J.B., K. Tsukimoto, O. Mathieu-Costello and R. Prediletto, Stress failure in pulmonary capillaries. *J. Appl. Physiol.* 70: 1731–1742, 1991.

Abstract: In the mammalian lung, alveolar gas and blood are separated by an extremely thin membrane, despite the fact that mechanical failure could be catastrophic for gas exchange. We raised the pulmonary capillary pressure in anesthetized rabbits until stress failure occurred. At capillary transmural pressures greater than or equal to 40 mm Hg, disruption of the capillary endothelium and alveolar epithelium was seen in some locations. The three principal forces acting on the capillary wall were analyzed. (1) Circumferential wall tension caused by the transmural pressure. This is approximately 25 dyn/cm (25 mN/m) at failure where the radius of curvature of the capillary is 5 microns. This tension is small, being comparable with the tension in the alveolar wall associated with lung elastic recoil. (2) Surface tension of the alveolar lining layer. This contributes support to the capillaries that bulge into the alveolar spaces at these high pressures. When protein leakage into the alveolar spaces occurs because of stress failure, the increase in surface tension caused by surfactant inhibition could be a powerful force preventing further failure. (3) Tension of the tissue elements in the alveolar wall associated with lung inflation. This may be negligible at normal lung volumes but considerable at high volumes. Whereas circumferential wall tension is low, capillary wall stress at failure is very high at approximately $8 \times 10(5)$ dyn/cm^2 ($8 \times 10(4)$ N/m^2) where the thickness is only 0.3 microns. This is approximately the same as the wall stress of the normal aorta, which is predominantly composed of collagen and elastin. The strength of the thin part of the capillary wall is probably attributable to the collagen IV of the basement membranes. The safety factor is apparently small when the capillary pressure is raised during heavy exercise. Stress failure causes increased permeability with protein leakage, or frank hemorrhage, and probably has a role in several types of lung disease.

Comment: This was the first of a series of articles on the structural changes in the walls of pulmonary capillaries that occur when they are exposed to increased capillary pressures, or the lung is inflated to a high volume. Stress failure of pulmonary capillaries is an important new concept in pulmonary physiology with many implications in lung disease.

272. West, J.B. and O. Mathieu-Costello, Strength of the pulmonary blood-gas barrier. *Resp. Physiol.* 88: 141–148, 1992.

Abstract: The mammalian pulmonary blood-gas barrier is well known to be extremely thin. For example, in the human lung, half of the area of the barrier (the ‘bulging’ part) has a thickness of only 0.2–0.4 micron. We show here that the barrier is also immensely strong. This is an essential requirement because the capillary wall stresses during heavy exercise become very large (about 7×10 (4) N/m^2 = 70 kPa) when capillary pressure increases to 30 mm Hg. Stress failure of the pulmonary capillary wall consistently occurs in experimental rabbit preparations at abnormally high pressures exceeding 40 mm Hg and may be the cause of bleeding into the lung in galloping racehorses. The great strength of the thin side of the blood-gas barrier can be attributed to the extracellular matrix, especially the type IV collagen which is predominantly located in the very thin lamina densa. The alveolar wall is therefore particularly vulnerable to injurious agents which attack type IV collagen such as autoantibodies in Goodpasture’s Syndrome and perhaps neutrophil elastase in emphysema. The combination of extreme thinness and great strength of the blood-gas barrier poses a unique design requirement.

Comment: Important paper providing the first evidence that the strength of the capillary wall comes from type IV collagen in the basement membranes.

284. Prisk, G.K., H.J.B. Guy, A.R. Elliott, R.A. Deutschman III and J.B. West, Pulmonary diffusing capacity, capillary blood volume and cardiac output during sustained microgravity. *J. Appl. Physiol.* 75: 15–26, 1993.

Abstract: We measured pulmonary diffusing capacity (DL), diffusing capacity per unit lung volume, pulmonary capillary blood volume (Vc), membrane diffusing capacity (Dm), pulmonary capillary blood flow or cardiac output (Qc), and cardiac stroke volume (SV) in four subjects exposed to 9 days of microgravity (weightlessness, 0 G). The same subjects were studied standing and supine numerous times preflight and in the week immediately after return from space. DL in microgravity was elevated (28%) compared with preflight standing values and was higher than preflight supine because of the elevation of both Vc (28%) and Dm (27%). The elevation in Vc was comparable to that measured supine in 1 G, but the increase in Dm was in sharp contrast to the supine value (which was unchanged). We postulate that, in 0 G, pulmonary capillary blood is evenly distributed throughout the lung, providing for uniform capillary filling, leading to an increase in the surface area available for diffusion. By contrast, in the supine 1-G state, the capillaries are less evenly filled, and although a similar increase in blood volume is observed, the corresponding increase in surface area does not occur. DL and its subdivisions showed no adaptive changes from the first measurement 24 h after the start of 0 G to 8 days later. Similarly, there were no trends in the postflight data, suggesting that the principal mechanism of these changes was gravitational. The increase in Dm suggests that subclinical pulmonary edema did not result from exposure to 0 G. Qc was modestly increased (18%) inflight and decreased (9%) post-flight compared with preflight standing. Compared with preflight standing, SV was increased 46% inflight and decreased 14% in the 1st week postflight. There were temporal changes in Qc and SV during 0 G, with the highest values recorded at the first measurement, 24 h into the flight. The lowest values of Qc and SV occurred on the day of return.

Comment: This is the first of a series of articles describing our measurements on astronauts in Spacelab. It was a rare privilege to make these sophisticated measurements in a completely new environment.

291. West, J.B., Krook's death by spontaneous combustion and the controversy between Dickens and Lewes: a physiologist's view. *The Dickensian* 90: 125–129, 1994.

Comment: This is one of my favorite articles. Krook was a character in Charles Dickens' novel Bleak House. He was a prodigious drinker of gin and died by bursting into flames. The prominent English physiologist, George Henry Lewes, took great exception to this literary absurdity, although Lewes' physiology was itself a little shaky. The article was published in a journal devoted solely to Dickens.

301. West, J.B., Oxygen enrichment of room air to relieve the hypoxia of high altitude. *Resp. Physiol.* 99: 225–232, 1995.

Abstract: Recently there has been increasing commercial activity at altitudes of 3500–6000 m. Examples include new mines in northern Chile at altitudes of about 4500 m. Because the workers come from sea level, intolerance of the high altitude is a major problem. This degree of hypoxia reduces work capacity, mental efficiency and sleep quality. One novel solution is to raise the PO2 of the room air by adding oxygen to the room ventilation. This is remarkably effective. For example, at altitudes of 4000–5000 m, increasing the O2 concentration by 1% (e.g. from 21 to 22%) reduces the equivalent altitude by about 300 m. Thus raising the O2 concentration by 5% at the new mines reduces the equivalent altitude to 3000 m which is easily tolerated. The introduction of oxygen concentrators (molecular sieve) which require only electrical power makes O2 enrichment feasible. The fire hazard is less than in air at sea level. Everybody now expects that the ventilation of a room will provide a comfortable temperature and humidity. Control of the oxygen concentration can be regarded as a further logical step in man's control of his environment.

Comment: This is the first article describing how oxygen can be added to the air of a room at high altitude thus reducing the physiological altitude. It was an important advance. The technique has been very successful in improving living and working conditions for people who need to be at high altitude.

316. Prisk, G.K., A.-M. Lauzon, S. Verbanck, A.R. Elliott, H.J.B. Guy, M. Paiva and J.B. West, Anomalous behavior of helium and sulfur hexafluoride during single-breath tests in sustained microgravity. *J. Appl. Physiol.* 80: 1126–1132, 1996.

Abstract: We performed single-breath wash-in tests for He and SF6 in four subjects exposed to 14 days of microgravity (microG) during the Spacelab flight Spacelab Life Sciences-2. Subjects inspired a vital capacity breath of 5% He-1.25% SF6-balance O2 and then exhaled to residual volume at 0.5l/s. The tests were also performed with a 10-s breath hold at the end of inspiration. Measurements were also made with the subjects standing and supine in 1 G. Phase III slope was measured after the dead-space washout and before the onset of airway closure. In all subjects in 1 G, whether standing or supine, phase III slope for SF6 was significantly steeper than that for He. However, in microG, the slopes became the same. Furthermore, after breath holding in microG, the SF6 slopes were significantly flatter than those for He. On return to 1 G, the changes were reversed, and there was no difference between preflight and postflight values. Because most of the phase III slope reflects events occurring in the acinar regions of the lung, the results suggest that microG causes conformational changes in the acini or changes in cardiogenic mixing in the lung periphery, but in either case the mechanism is unclear.

Comment: This is a report of the strange behavior of inhaled gases of very different molecular weights in microgravity. The results suggest different diffusion rates but these rates should not be affected by gravity. The behavior has not yet been satisfactorily explained.

326. Hopkins, S.R., R.B. Schoene, W.R. Henderson, R.G. Spragg, T.R. Martin and J.B. West, Intense exercise impairs the integrity of the pulmonary blood-gas barrier in elite athletes. *Am. J. Resp. Crit. Care Med.* 155: 1090–1094, 1997

Abstract: The blood-gas barrier must be very thin to allow gas exchange and it is therefore subjected to high mechanical stresses when the capillary pressure rises. In some animals, such as the thoroughbred race-horse during intense exercise, the stresses are so large that the capillaries fail and bleeding occurs. We tested the hypothesis that, in elite human athletes, the high capillary pressure that occurs during severe exercise alters the structure and function of the blood-gas barrier. We performed bronchoalveolar lavage (BAL) in six healthy athletes, who had a history suggestive of lung bleeding, 1 h after a 7-min cycling race simulation and four normal sedentary control subjects who did not exercise before BAL. The athletes had higher ($p < 0.05$) concentrations of red blood cells ($0.51 \times 10(5)$ versus $0.01 \times 10(5).ml^{-1}$), total protein (128.0 versus 94.1 micrograms/ml), albumin (65.6 versus 53.0 micrograms/ml), and leukotriene B4 (LTB4) (243 versus 0 pg/ml) in BAL fluid than control subjects. The proportion of neutrophils was similar in athletes and control subjects but the proportion of lymphocytes in BAL fluid was reduced ($p < 0.05$). There were no differences in levels of surfactant apoprotein A, tumor necrosis factor bioactivity, lipopolysaccharide, or interleukin-8 (IL-8) between groups. These results show that brief intense exercise in athletes with a history suggestive of lung bleeding alters blood-gas barrier function resulting in higher concentrations of red cells and protein in BAL fluid. The lack of activation of proinflammatory pathways (except LTB4) in the airspaces supports the hypothesis that the mechanism for altered blood-gas barrier function is mechanical stress.

Comment: This was an important study showing that maximal exercise in elite athletes apparently causes changes in the structure of the walls of their pulmonary capillaries.

329. Berg, J.T., Z. Fu, E.C. Breen, H.-C. Tran, O. Mathieu-Costello and J.B. West, High lung inflation increases mRNA levels of ECM components and growth factors in lung parenchyma. *J. Appl. Physiol.* 83: 120–128, 1997.

Abstract: Remodeling of pulmonary capillaries occurs after chronic increases in capillary pressure (e.g., mitral stenosis). Also, remodeling of pulmonary arteries begins within 4 h of increased wall stress and is endothelium dependent. We have previously shown that high lung inflation increases wall stress in pulmonary capillaries. This study was designed to determine whether high lung inflation induces remodeling of the extracellular matrix (ECM) in lung parenchyma. Open-chest rabbits were ventilated for 4 h with 9-cmH2O positive end-expiratory pressure (PEEP) on one lung and 1-cmH2O PEEP on the other

(High-PEEP group), or with 2-cmH2O PEEP on both lungs (Low-PEEP group). An additional untreated control group was also included. We found increased levels of mRNA in both lungs of High-PEEP rabbits (compared with both the Low-PEEP and untreated groups) for α1(III) and α2(IV) procollagen, fibronectin, basic fibroblast growth factor, and transforming growth factor-β1. In contrast, α2(I) procollagen and vascular endothelial growth factor mRNA levels were not changed. We conclude that high lung inflation for 4 h increases mRNA levels of ECM components and growth factors in lung parenchyma.

Comment: This was our first study of changes in gene expression apparently caused by increased stress in the blood-gas barrier. This is important because of evidence that remodeling of the barrier occurs when the capillary pressure rises in disease, for example, mitral stenosis.

335. Darquenne, C., M. Paiva, J.B. West and G.K. Prisk, Effect of microgravity and hypergravity on deposition of 0.5- to 3-μm-diameter aerosol in the human lung. *J. Appl. Physiol.* 83: 2029–2036, 1997.

Abstract: We measured intrapulmonary deposition of 0.5-, 1-, 2-, and 3-μm-diameter particles in four subjects on the ground (1 G) and during parabolic flights both in microgravity (μG) and at ~1.6 G. Subjects breathed aerosols at a constant flow rate (0.4 l/s) and tidal volume (0.75 L). At 1 G and ~1.6 G, deposition increased with increasing particle size. In μG, differences in deposition as a function of particle size were almost abolished. Deposition was a nearly linear function of the G level for 2- and 3-μm-diameter particles, whereas for 0.5- and 1.0-μm-diameter particles, deposition increased less between μG and 1 G than between 1 G and ~1.6 G. Comparison with numerical predictions showed good agreement for 1-, 2-, and 3-μm-diameter particles at 1 and ~1.6 G, whereas the model consistently underestimated deposition in μG. The higher deposition observed in μG compared with model predictions might be explained by a larger deposition by diffusion because of a higher alveolar concentration of aerosol in μG and to the nonreversibility of the flow, causing additional mixing of the aerosols.

Comment: This study shows that microgravity alters the deposition of aerosol in the lung. The study is important from a theoretical point of view, but in addition NASA is concerned about possible hazards of toxic dust on Mars and the Moon.

354. West, J.B., Eugene Gippenreiter (1927–1997): a personal appreciation. *Aviakosmicheskaya I Ekologicheskaya Meditsina*, Supplement 2: 81–83 (Russian), 83–84 (English), 1998.

Comment: I felt very privileged to be asked to write this appreciation for a Russian journal. Eugene Gippenreiter was a Russian friend whom I first met when I was president of the American Physiological Society, and was invited to visit the Physiological Society of the Soviet Union in 1985 (see publication #217). Gippenreiter was a prominent mountaineer, and he was also connected with the Soviet space program. Subsequently our two families including his son, and my wife, son and daughter took part in a 2 week trek in southern Russia near Mt

Elbrus, and a very remote part of Georgia. This was an unforgettable experience. He also visited us in La Jolla.

384. West, J.B., Snorkel breathing in the elephant explains the unique anatomy of its pleura. *Resp. Physiol.* 126: 1–8, 2001

Abstract: It has been known for over 300 years that the anatomy of the elephant lung is unique among mammals in that the pleural cavity is obliterated by connective tissue. However no satisfactory explanation has been advanced. Recent studies suggest that the elephant has an aquatic ancestry and the trunk may have developed for snorkeling. In addition, the modern day elephant is the only mammal that can remain submerged far below the surface of the water while snorkeling. The resulting differences of pressures within the thorax mean that the small blood vessels of the pleura are in great danger of rupturing or causing severe edema. The same distribution of pressures occurs when the animal raises water inside its trunk prior to drinking although in this case the pressure differences are relatively short-lived. Evolution has provided a remarkable solution to this problem by replacing the normally delicate parietal and visceral pleurae by dense connective tissue, and separating the two pleurae by loose connective tissue to allow some sliding movement.

Comment: This is one of my most original and colorful studies, and so far seems to have stood the test of time.

387. West, J.B., Historical aspects of the Early Soviet/Russian manned space program. *J. Appl. Physiol.* 91: 1501–1511, 2001.

Abstract: Human spaceflight was one of the great physiological and engineering triumphs of the 20th century. Although the history of the United States manned space program is well known, the Soviet program was shrouded in secrecy until recently. Konstantin Edvardovich Tsiolkovsky (1857–1935) was an extraordinary Russian visionary who made remarkable predictions about space travel in the late nineteenth century. Sergei Pavlovich Korolev (1907–1966) was the brilliant "Chief Designer" who was responsible for many of the Soviet firsts, including the first artificial satellite and the first human being in space. The dramatic flight of Sputnik 1 was followed within a month by the launch of the dog Laika, the first living creature in space. Remarkably, the engineering work for this payload was all done in less than 4 wk. Korolev's greatest triumph was the flight of Yuri Alekseyevich Gagarin (1934–1968) on April 12, 1961. Another extraordinary feat was the first extravehicular activity by Aleksei Arkhipovich Leonov (1934–) using a flexible airlock that emphasized the entrepreneurial attitude of the Soviet engineers. By the mid-1960s, the Soviet program was overtaken by the United States program and attempts to launch a manned mission to the Moon failed. However, the early Soviet manned space program has a preeminent place in the history of space physiology

Comment: This is mainly about the contributions of Sergei Korolev who was an outstanding Soviet spaceflight pioneer. His career makes great reading.

401. West, J.B., Highest permanent human habitation. *High Alt. Med. Biol.* 3: 401–407, 2002.

Abstract: The aim of this analysis was to determine the altitude of the highest permanent human habitation in the hope that this will throw some light on what determines the highest altitude that a community can tolerate indefinitely. A number of places where people have lived at very high altitudes for long periods of time are reviewed. Individuals have lived for as long as 2 year at an altitude of 5950 m, and there was a miner's camp at 5300 m for several years. The highest permanently inhabited town in the world at the present time appears to be La Rinconada, a mining village of over 7000 people in southern Peru at an altitude of up to 5100 m, which has been in existence for over 40 year. The altitude of the highest permanent human habitation is determined partly by economic factors, rather than solely by human tolerance to hypoxia.

Comment: When I started this project, several Peruvians claimed that the town was considerably higher, but I was able to find a map with altitude contours prepared by the Peruvian military.

445. West, J.B., R.R. Watson and Z. Fu. The human lung: did evolution get it wrong? Eur. Resp. J. 29: 11–17, 2007.

Abstract: Some 300 million years ago, the ancestors of modern reptiles emerged from water and were committed to air breathing. They were exothermic and incapable of sustained levels of high physical activity. But from them evolved the two great classes of vertebrates with high levels of maximal oxygen consumption: the mammals and birds. A remarkable feature of these two divergent evolutionary lines is that, although the physiology of many organ systems shows many similarities, the lungs are radically different. A major difference is that the ventilation of the gas-exchanging tissue has a flow-through pattern in the bird but is reciprocating in the mammal. The result is that mammals have a reduced alveolar and arterial oxygen tension, a potential for uneven ventilation, and relatively large terminal air spaces. This in turn means that the pulmonary capillaries are poorly supported compared with the bird. The result is that the pulmonary capillaries in the bird have much thinner and more uniform walls, with more efficient gas exchange. Other advantages of the bird lung are that it utilises a more efficient cross-current pattern of gas-exchange, and the bird has separated the ventilatory and gas exchange functions. From a structure–function standpoint, the bird lung is superior.

Comment: A somewhat lighthearted article pointing out that the avian lung has many advantages compared with the mammalian lung. A major advantage is that the bird had separated the ventilatory and gas exchange functions.

462. West, J.B. A new approach to very-high-altitude land travel: the train to Lhasa, Tibet. *Ann. Int. Med.* 149:898–900, 2008.

Abstract: The train from Golmud in Qinghai province in China to Lhasa,Tibet, travels for more than 14 hours at an average altitude of 4500 m, the highest point being 5072 m (16,600 ft). Therefore, passengers are potentially exposed to very severe sustained hypoxia. The enterprising solution is to increase the oxygen concentration of the air in the train from 21% to between 24% and 25% by the use of oxygen generators in each passenger car. By increasing the oxygen concentration to 25%, the equivalent altitude is reduced by about 1200 m. The engineering challenge of oxygenating the whole train is immense but has been successfully accomplished. This is an extraordinary innovation in the high-altitude transportation.

Comment: The engineering is outstanding. However the principle is the same as that set out by me in publication #301 above. I gave a lecture in Lhasa on this topic a few years before the train appeared and like to think that I made a contribution.

499. West, J.B. A strategy for in-flight measurements of physiology of pilots of high-performance fighter aircraft. *J Appl Physiol.* 115:145–149, 2013.

Abstract: Some pilots flying modern high-performance fighter aircraft develop "hypoxia-like" incidents characterized by short periods of confusion and cognitive impairment. The problem is serious and recently led to the grounding of a fleet of aircraft. Extensive discussions of the incidents have taken place but some people believe that there is inadequate data to determine the cause. There is a tremendous disconnect between what is known about the function of the aircraft and the function of the pilot. This paper describes a plan for measuring the inspired and expired Po2 and Pco2 in the pilot's mask, the inspiratory flow rate, and pressure in the mask. A critically important requirement is that the interference with the function of the pilot is minimal. Although extensive physiological measurements were previously made on pilots in ground-based experiments such as rapid decompression in an altitude chamber and increased acceleration on a centrifuge, in-flight measurements of gas exchange have not been possible until now primarily because of the lack of suitable equipment. The present paper shows how the recent availability of small, rapidly responding oxygen and carbon dioxide analyzers make sophisticated in-flight measurements feasible. The added information has the potential of greatly improving our knowledge of pilot physiology, which could lead to an explanation for the incidents.

Comment: The fact that so little is known about the function of the pilot in these high-performance aircraft is bizarre. Kim Prisk and I are currently developing equipment that we hope will throw light on this important problem.

MIX
Papier aus verantwortungsvollen Quellen
Paper from responsible sources
FSC® C105338

If you have any concerns about our products,
you can contact us on
ProductSafety@springernature.com

In case Publisher is established outside the EU,
the EU authorized representative is:
Springer Nature Customer Service Center GmbH
Europaplatz 3, 69115 Heidelberg, Germany

Printed by Libri Plureos GmbH
in Hamburg, Germany